SCHAUM'S
EASY OUTLINES

Mathematical Handbook of Formulas and Tables

Online Diagnostic Test

Go to **Schaums.com** to launch the Schaum's Diagnostic Test.

This convenient application provides a 30-question multiple-choice test that will pinpoint areas of strength and weakness to help you focus your study. Questions cover all aspects of beginning chemistry, and the correct answers are explained in full. With a question-bank that rotates daily, the Schaum's Online Test also allows you to check your progress and readiness for final exams.

Other titles featured in Schaum's Online Diagnostic Test:

Schaum's Easy Outlines: Biochemistry
Schaum's Easy Outlines: Linear Algebra
Schaum's Easy Outlines: Molecular and Cell Biology
Schaum's Easy Outlines: Probability and Statistics
Schaum's Easy Outlines: Principles of Accounting
Schaum's Easy Outlines: Logic
Schaum's Easy Outlines: Bookkeeping and Accounting
Schaum's Easy Outlines: College Mathematics
Schaum's Easy Outlines: Applied Physics
Schaum's Easy Outlines: College Physics
Schaum's Easy Outlines: Differential Equations
Schaum's Easy Outlines: Basic Electricity
Schaum's Easy Outlines: Spanish, 2nd Edition
Schaum's Easy Outlines: French, 2nd Edition
Schaum's Easy Outlines: German, 2nd Edition
Schaum's Easy Outlines: Italian, 2nd Edition
Schaum's Easy Outlines: Writing and Grammar, 2nd Edition
Schaum's Easy Outlines: Geometry, 2nd Edition
Schaum's Easy Outlines: Calculus, 2nd Edition
Schaum's Easy Outlines: Statistics, 2nd Edition
Schaum's Easy Outlines: Elementary Algebra, 2nd Edition
Schaum's Easy Outlines: College Algebra, 2nd Edition
Schaum's Easy Outlines: Biology, 2nd Edition
Schaum's Easy Outlines: Human Anatomy and Physiology, 2nd Edition
Schaum's Easy Outlines: Organic Chemistry, 2nd Edition
Schaum's Easy Outlines: College Chemistry, 2nd Edition

Mathematical Handbook of Formulas and Tables

Murray R. Spiegel, Ph.D.
and Jon Liu, Ph.D.

Abridgement Editor:
George J. Hademenos, Ph.D.

New York   Chicago   San Francisco   Lisbon   London   Madrid   Mexico City
Milan   New Delhi   San Juan   Seoul   Singapore   Sydney   Toronto

The *McGraw-Hill* Companies

The late **MURRAY R. SPIEGEL** an M.S. degree in Physics and a Ph.D. in Mathematics from Cornell University. He taught at Harvard, Columbia, and Rensselaer Polytechnic Institute, and worked at Oak Ridge as a mathematical consultant.

JON LIU is a professor of Mathematics at Temple University. He received his Ph.D. from the University of California and held visiting positions at New York University, Princeton University, and the University of California at Berkeley.

GEORGE J. HADEMENOS has taught at the University of Dallas and has performed research at the University of California at Los Angeles and the University of Massachusetts Medical Center. He earned a B.S. degree from Angelo State University and a Ph.D. in Physics from the University of Texas at Dallas.

2 3 4 5 6 7 8 9 10 11 12 13 14 15   DOC/DOC   1 9 8 7 6 5 4 3

ISBN   978-00-7177747-6
MHID   0-07-177747-4

McGraw-Hill books are available at special quantity discounts to use as premiums and sales promotions or for use in corporate training programs. To contact a representative, please e-mail us at bulksales@mcgraw-hill.com.

This book is printed on acid-free paper.

Contents

PART A

FORMULAS

Section I
ELEMENTARY CONSTANTS, PRODUCTS, FORMULAS

1. Special Constants

1.1 $\pi = 3.14159\ 26535\ 89793\ 23846\ 2643\dots$

1.2 $e = 2.71828\ 18284\ 59045\ 23536\ 0287\dots = \lim\limits_{n\to\infty}\left(1 + \dfrac{1}{n}\right)^n$
      = natural base of logarithms

1.3 $\sqrt{2} = 1.41421\ 35623\ 73095\ 0488\dots$

1.4 $\sqrt{3} = 1.73205\ 08075\ 68877\ 2935\dots$

1.5 $\sqrt[3]{2} = 1.25992\ 1050\dots$

1.6 $\sqrt[3]{2} = 1.25992\ 1050\dots$

1.7 $e^\pi = 23.14069\ 26327\ 79269\ 006\dots$

1.8 $\pi^e = 22.45915\ 77183\ 61045\ 47342\ 715\dots$

1.9 $e^e = 15.15426\ 22414\ 79264\ 190\dots$

3

1.10 $\log_{10} 2 = 0.30102\ 99956\ 63981\ 19521\ 37389\ldots$

1.11 $\log_{10} 3 = 0.47712\ 12547\ 19662\ 43729\ 50279\ldots$

1.12 $\log_{10} e = 0.43429\ 44819\ 03251\ 82765\ldots$

1.13 $\log_{10} \pi = 0.49714\ 98726\ 94133\ 85435\ 12683\ldots$

1.14 $\log_e 10 = \ln 10 = 2.30258\ 50929\ 94045\ 68401\ 7991\ldots$

1.15 $\log_e 2 = \ln 2 = 0.69314\ 71805\ 59945\ 30941\ 7232\ldots$

1.16 $\log_e 3 = \ln 3 = 1.09861\ 22886\ 68109\ 69139\ 5245\ldots$

1.17 $\gamma = 0.57721\ 56649\ 01532\ 86060\ 6512\ldots = $ *Euler's constant*

$$= \lim_{n \to \infty} \left(1 + \frac{1}{2} + \frac{1}{3} + \cdots + \frac{1}{n} - \ln n \right)$$

1.18 $e^{\gamma} = 1.78107\ 24179\ 90197\ 9852\ldots$ [see 1.17]

1.19 $\sqrt{e} = 1.64872\ 12707\ 00128\ 1468\ldots$

1.20 $\sqrt{\pi} = 1.77245\ 38509\ 05516\ 02729\ 8167\ldots$

1.21 $1 \text{ radian} = 180°/\pi = 57.29577\ 95130\ 8232\ldots°$

1.22 $1° = \pi/180 \text{ radians} = 0.01745\ 32925\ 19943\ 29576\ 92\ldots \text{ radians}$

2. Special Products and Factors

2.1 $(x + y)^2 = x^2 + 2xy + y^2$

2.2 $(x - y)^2 = x^2 - 2xy + y^2$

2.3 $(x + y)^3 = x^3 + 3x^2 y + 3xy^2 + y^3$

2.4 $(x - y)^3 = x^3 - 3x^2 y + 3xy^2 - y^3$

2.5 $(x + y)^4 = x^4 + 4x^3 y + 6x^2 y^2 + 4xy^3 + y^4$

2.6 $(x-y)^4 = x^4 - 4x^3y + 6x^2y^2 - 4xy^3 + y^4$

2.7 $(x+y)^5 = x^5 + 5x^4y + 10x^3y^2 + 10x^2y^3 + 5xy^4 + y^5$

2.8 $(x-y)^5 = x^5 - 5x^4y + 10x^3y^2 - 10x^2y^3 + 5xy^4 - y^5$

2.9 $(x+y)^6 = x^6 + 6x^5y + 15x^4y^2 + 20x^3y^3 + 15x^2y^4 + 6xy^5 + y^6$

2.10 $(x-y)^6 = x^6 - 6x^5y + 15x^4y^2 - 20x^3y^3 + 15x^2y^4 - 6xy^5 + y^6$

    The results 2.1 to 2.10 above are special cases of the binomial formula [see 3.3].

2.11 $x^2 - y^2 = (x-y)(x+y)$

2.12 $x^3 - y^3 = (x-y)(x^2 + xy + y^2)$

2.13 $x^3 + y^3 = (x+y)(x^2 - xy + y^2)$

2.14 $x^4 - y^4 = (x-y)(x+y)(x^2 + y^2)$

2.15 $x^5 - y^5 = (x-y)(x^4 + x^3y + x^2y^2 + xy^3 + y^4)$

2.16 $x^5 + y^5 = (x+y)(x^4 - x^3y + x^2y^2 - xy^3 + y^4)$

2.17 $x^6 - y^6 = (x-y)(x+y)(x^2 + xy + y^2)(x^2 - xy + y^2)$

2.18 $x^4 + x^2y^2 + y^4 = (x^2 + xy + y^2)(x^2 - xy + y^2)$

2.19 $x^4 + 4y^4 = (x^2 + 2xy + 2y^2)(x^2 - 2xy + 2y^2)$

    Some generalizations of the above are given by the following results where n is a positive integer.

2.20

$$x^{2n+1} - y^{2n+1} = (x-y)(x^{2n} + x^{2n-1}y + x^{2n-2}y^2 + \cdots + y^{2n})$$
$$= (x-y)\left(x^2 - 2xy\cos\frac{2\pi}{2n+1} + y^2\right)\left(x^2 - 2xy\cos\frac{4\pi}{2n+1} + y^2\right)$$
$$\cdots\left(x^2 - 2xy\cos\frac{2n\pi}{2n+1} + y^2\right)$$

2.21

$$x^{2n+1} + y^{2n+1} = (x+y)(x^{2n} - x^{2n-1}y + x^{2n-2}y^2 - \cdots + y^{2n})$$

$$= (x+y)\left(x^2 + 2xy\cos\frac{2\pi}{2n+1} + y^2\right)\left(x^2 + 2xy\cos\frac{4\pi}{2n+1} + y^2\right)$$

$$\cdots\left(x^2 + 2xy\cos\frac{2n\pi}{2n+1} + y^2\right)$$

2.22

$$x^{2n} - y^{2n} = (x-y)(x+y)(x^{n-1} + x^{n-2}y + x^{n-3}y^2 + \cdots)(x^{n-1} - x^{n-2}y + x^{n-3}y^2 - \cdots)$$

$$= (x-y)(x+y)\left(x^2 - 2xy\cos\frac{\pi}{n} + y^2\right)\left(x^2 - 2xy\cos\frac{2\pi}{n} + y^2\right)$$

$$\cdots\left(x^2 - 2xy\cos\frac{(n-1)\pi}{n} + y^2\right)$$

2.23 $\quad x^{2n} + y^{2n} = \left(x^2 + 2xy\cos\dfrac{\pi}{2n} + y^2\right)\left(x^2 + 2xy\cos\dfrac{3\pi}{2n} + y^2\right)$

$$\cdots\left(x^2 + 2xy\cos\frac{(2n-1)\pi}{2n} + y^2\right)$$

3. The Binomial Formula

Factorial *n*

For $n = 1, 2, 3, \ldots,$ *factorial n* or *n factorial* is denoted and defined by:

3.1 $\quad n! = 1 \cdot 2 \cdot 3 \cdots\cdots n$

Zero factorial is defined by

3.2 $\quad 0! = 1$

n factorial can be defined recursively by

$0! = 1 \quad$ and $\quad n! = n \bullet (n-1)!$

Example 3.1.

$4! = 1 \bullet 2 \bullet 3 \bullet 4 = 24$
$5! = 1 \bullet 2 \bullet 3 \bullet 4 \bullet 5 = 5 \bullet 4! = 5(24) = 120$
$6! = 1 \bullet 2 \bullet 3 \bullet 4 \bullet 5 \bullet 6 = 6 \bullet 5! = 6(120) = 720$

Binomial Formula for Positive Integral n

For $n = 1, 2, 3, \ldots$

3.3

$$(x + y)^n = x^n + nx^{n-1}y + \frac{n(n - 1)}{2!} x^{n-2}y^2 + \frac{n(n - 1)(n - 2)}{3!} x^{n-3}y^3 + \cdots + y^n$$

This is called the *binomial formula*. It can be extended to other values of n, and also to an infinite series.

Example 3.2.

$(a - 2b)^4 = a^4 + 4a^3(-2b) + 6a^2(-2b)^2 + 4a(-2b)^3 + (-2b)^4 = a^4 - 8a^3b + 24a^2b^2 - 32ab^3 + 16b^4$

Here $x = a$ and $y = -2b$.

The binomial formula can be simplified, as shown in Figure 3-1.

$$(a + b)^0 = 1$$
$$(a + b)^1 = a + b$$
$$(a + b)^2 = a^2 + 2ab + b^2$$
$$(a + b)^3 = a^3 + 3a^2b + 3ab^2 + b^3$$
$$(a + b)^4 = a^4 + 4a^3b + 6a^2b^2 + 4ab^3 + b^4$$
$$(a + b)^5 = a^5 + 5a^4b + 10a^3b^2 + 10a^2b^3 + 5ab^4 + b^5$$
$$(a + b)^6 = a^6 + 6a^5b + 15a^4b^2 + 20a^3b^3 + 15a^2b^4 + 6ab^5 + b^6$$

. . . .

Figure 3-1

Binomial Coefficients

Formula 3.3 can be rewritten in the form

3.4 $$(x+y)^n = x^n + \binom{n}{1}x^{n-1}y + \binom{n}{2}x^{n-2}y^2 + \binom{n}{3}x^{n-3}y^3 + \cdots + \binom{n}{n}y^n$$

where the coefficients, called *binomial coefficients*, are given by:

3.5 $$\binom{n}{k} = \frac{n(n-1)(n-2)\cdots(n-k+1)}{k!} = \frac{n!}{k!(n-k)!} = \binom{n}{n-k}$$

Example 3.3.

$$\binom{9}{4} = \frac{9\cdot 8\cdot 7\cdot 6}{1\cdot 2\cdot 3\cdot 4} = 126, \quad \binom{12}{5} = \frac{12\cdot 11\cdot 10\cdot 9\cdot 8}{1\cdot 2\cdot 3\cdot 4\cdot 5} = 792,$$

$$\binom{10}{7} = \binom{10}{3} = \frac{10\cdot 9\cdot 8}{1\cdot 2\cdot 3} = 120$$

The binomial coefficients may be arranged in a triangular array of numbers, called *Pascal's triangle*, as shown in Figure 3-2.

```
            1
          1   1
        1   2   1
      1   3   3   1
    1   4   6   4   1
  1   5  10  10   5   1
1   6  15  20  15   6   1
. . . . . . . . . . . . . .
```

Figure 3-2

The triangle has the following two properties:

(1) The first and last number in each row is 1.

(2) Every other number in the array can be obtained by adding the two numbers appearing directly above it. Property (2) may be stated as follows:

3.6 $\binom{n}{k} + \binom{n}{k+1} = \binom{n+1}{k+1}$

Properties of Binomial Coefficients

The following lists additional properties of the binomial coefficients:

3.7 $\binom{n}{0} + \binom{n}{1} + \binom{n}{2} + \cdots + \binom{n}{n} = 2^n$

3.8 $\binom{n}{0} - \binom{n}{1} + \binom{n}{2} - \cdots(-1)^n\binom{n}{n} = 0$

3.9 $\binom{n}{n} + \binom{n+1}{n} + \binom{n+2}{n} + \cdots + \binom{n+m}{n} = \binom{n+m+1}{n+1}$

3.10 $\binom{n}{0} + \binom{n}{2} + \binom{n}{4} + \cdots = 2^{n-1}$

3.11 $\binom{n}{1} + \binom{n}{3} + \binom{n}{5} + \cdots = 2^{n-1}$

3.12 $\binom{n}{0}^2 + \binom{n}{1}^2 + \binom{n}{2}^2 + \cdots + \binom{n}{n}^2 = \binom{2n}{n}$

3.13 $\binom{m}{0}\binom{n}{p} + \binom{m}{1}\binom{n}{p-1} + \cdots + \binom{m}{p}\binom{n}{0} = \binom{m+n}{p}$

3.14 $(1)\binom{n}{1} + (2)\binom{n}{2} + (3)\binom{n}{3} + \cdots + (n)\binom{n}{n} = n2^{n-1}$

3.15 $(1)\binom{n}{1} - (2)\binom{n}{2} + (3)\binom{n}{3} - \cdots(-1)^{n+1}(n)\binom{n}{n} = 0$

Multinomial Formula

Let $n_1, n_2, \ldots, n_r$ be nonnegative integers such that $n_1 + n_2 + \ldots + n_r = n$. Then the following expression, called a *multinomial coefficient*, is defined as follows:

3.16 $\begin{pmatrix} n \\ n_1, n_2, \ldots, n_r \end{pmatrix} = \dfrac{n!}{n_1! n_2! \cdots n_r!}$

The name *multinomial coefficient* comes from the following formula:

3.17 $(x_1 + x_2 + \cdots + x_p)^n = \displaystyle\sum \begin{pmatrix} n \\ n_1, n_2, \ldots, n_r \end{pmatrix} x_1^{n_1} x_2^{n_2} \cdots x_r^{n_r}$

where the sum, denoted by Σ, is taken over all possible multinomial coefficients.

4. Complex Numbers

Definitions Involving Complex Numbers

A complex number z is generally written in the form

$z = a + bi$

where a and b are real numbers and i, called the *imaginary unit*, has the property that $i^2 = -1$. The real numbers a and b are called the *real* and *imaginary* parts of $z = a + bi$, respectively.

The *complex conjugate* of z is denoted by $\overline{z}$; it is defined by:

$\overline{(a + bi)} = a - bi$

Thus $a + bi$ and $a - bi$ are conjugates of each other.

Equality of Complex Numbers

4.1 $a + bi = c + di$ **if and only if** $a = c$ **and** $b = d$

Arithmetic of Complex Numbers

Formulas for the addition, subtraction, multiplication, and division of complex numbers follow:

4.2 $(a + bi) + (c + di) = (a + c) + (b + d)i$

4.3 $(a + bi) - (c + di) = (a - c) + (b - d)i$

4.4 $(a + bi)(c + di) = (ac - bd) + (ad + bc)i$

4.5 $\dfrac{a + bi}{c + di} = \dfrac{a + bi}{c + di} \cdot \dfrac{c - di}{c - di} = \dfrac{ac + bd}{c^2 + d^2} + \left(\dfrac{bc - ad}{c^2 + d^2} \right) i$

Note that the above operations are obtained by using the ordinary rules of algebra and replacing i^2 by -1 wherever it occurs.

Example 4.1.

Suppose $z = 2 + 3i$ and $w = 5 - 2i$. Then

$$z + w = (2 + 3i) + (5 - 2i) = 2 + 5 + 3i - 2i = 7 + i$$

$$zw = (2 + 3i)(5 - 2i) = 10 + 15i - 4i - 6i^2 = 16 + 11i$$

$$\overline{z} = \overline{(2 + 3i)} = 2 - 3i \ \text{ and } \ \overline{w} = \overline{(5 - 2i)} = 5 + 2i$$

$$\frac{w}{z} = \frac{5 - 2i}{2 + 3i} = \frac{(5 - 2i)(2 - 3i)}{(2 + 3i)(2 - 3i)} = \frac{4 - 19i}{13} = \frac{4}{13} - \frac{19}{13}i$$

Complex Plane

Real numbers can be represented by the points on a line, called the *real line*, and similarly, complex numbers can be represented by points in the plane, called the *Argand diagram* or the *complex plane*. Specifically, we let the point (a,b) in the plane represent the complex number $z = a + bi$. For example, the point P in Figure 4-1 represents the complex number $z = -3 + 4i$.

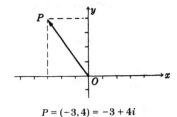

$$P = (-3, 4) = -3 + 4i$$

Figure 4-1

The complex number can also be interpreted as a vector from the origin O to the point P, as shown in Figure 4-2.

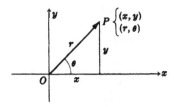

Figure 4-2

The *absolute value* of a complex number $z = a + bi$, written $|z|$, is defined as follows:

4.6 $\quad |z| = \sqrt{a^2 + b^2} = \sqrt{z\bar{z}}$

We note $|z|$ is the distance from the origin O to the point z in the complex plane.

Polar Form of Complex Numbers

The point P in Fig. 4-2 with coordinates (x,y) represents the complex number $z = x + iy$. The point P can also be represented by *polar coordinates* (r, θ). Since $x = r \cos \theta$ and $y = r \sin \theta$, we have

4.7 $\quad z = x + iy = r(\cos \theta + i \sin \theta)$

called the *polar form* of the complex number. We often call $r = |z|$ the *modulus* and θ the *amplitude* of $z = x + iy$.

Multiplication and Division of Complex Numbers in Polar Form
4.8
$$[r_1(\cos \theta_1 + i \sin \theta_1)][r_2(\cos \theta_2 + i \sin \theta_2)] = r_1 r_2[\cos (\theta_1 + \theta_2) + i \sin (\theta_1 + \theta_2)]$$

4.9 $\quad \dfrac{r_1(\cos \theta_1 + i \sin \theta_1)}{r_2(\cos \theta_2 + i \sin \theta_2)} = \dfrac{r_1}{r_2}[\cos (\theta_1 - \theta_2) + i \sin (\theta_1 - \theta_2)]$

De Moivre's Theorem

For any real number p, De Moivre's theorem states that:

4.10 $[r(\cos\theta + i\sin\theta)]^p = r^p(\cos p\theta + i\sin p\theta)$

Roots of Complex Numbers

Let $p = 1/n$ where n is any positive integer. Then 4.10 can be written

4.11 $[r(\cos\theta + i\sin\theta)]^{1/n} = r^{1/n}\left(\cos\dfrac{\theta + 2k\pi}{n} + i\sin\dfrac{\theta + 2k\pi}{n}\right)$

where k is any integer. From this formula, all the n nth roots of a complex number can be obtained by putting $k = 0, 1, 2, ..., n - 1$.

5. Solutions of Algebraic Equations

Quadratic Equation: $ax^2 + bx + c = 0$

5.1 Solutions: $x = \dfrac{-b \pm \sqrt{b^2 - 4ac}}{2a}$

If a, b, c are real and if $D = b^2 - 4ac$ is the *discriminant*, then the roots are

(1) real and unreal if $D > 0$
(2) real and equal if $D = 0$
(3) complex conjugate if $D < 0$

5.2 If x_1, x_2 are the roots, then $x_1 + x_2 = -b/a$ and $x_1 x_2 = c/a$.

Cubic Equation: $x^3 + a_1 x^2 + a_2 x + a_3 = 0$

Let

$$Q = \frac{3a_2 - a_1^2}{9}, \quad R = \frac{9a_1 a_2 - 27a_3 - 2a_1^3}{54},$$

$$S = \sqrt[3]{R + \sqrt{Q^3 + R^2}}, \quad T = \sqrt[3]{R - \sqrt{Q^3 + R^2}}$$

where $ST = -Q$.

5.3 Solutions:
$$
\begin{cases}
x_1 = S + T - \tfrac{1}{3}a_1 \\
x_2 = -\tfrac{1}{2}(S + T) - \tfrac{1}{3}a_1 + \tfrac{1}{2}i\sqrt{3}(S - T) \\
x_3 = -\tfrac{1}{2}(S + T) - \tfrac{1}{3}a_1 - \tfrac{1}{2}i\sqrt{3}(S - T)
\end{cases}
$$

If a_1, a_2, a_3 are real and if $D = Q^3 + R^2$ is the *discriminant*, then

(1) one root is real and two complex conjugate if $D > 0$
(2) all roots are real and at least two are equal if $D = 0$
(3) all roots are real and unequal if $D < 0$

If $D < 0$, computation is simplified by use of trigonometry.

5.4 Solutions if $D < 0$:
$$
\begin{cases}
x_1 = 2\sqrt{-Q} \cos (\tfrac{1}{3}\theta) - \tfrac{1}{3}a_1 \\
x_2 = 2\sqrt{-Q} \cos (\tfrac{1}{3}\theta + 120°) - \tfrac{1}{3}a_1 \\
x_3 = 2\sqrt{-Q} \cos (\tfrac{1}{3}\theta + 240°) - \tfrac{1}{3}a_1
\end{cases}
$$

where $\cos \theta = R/\sqrt{-Q^3}$

5.5 $\quad x_1 + x_2 + x_3 = -a_1, \quad x_1x_2 + x_2x_3 + x_3x_1 = a_2, \quad x_1x_2x_3 = -a_3$

where x_1, x_2, x_3 are the three roots.

Quartic Equation: $x^4 + a_1x^3 + a_2 x^2 + a_3 x + a_4 = 0$

Let y_1 be a real root of the following cubic equation:

5.6 $\quad y^3 - a_2y^2 + (a_1a_3 - 4a_4)y + (4a_2a_4 - a_3^2 - a_1^2a_4) = 0$

The four roots of the quartic equation are the four roots of the following equation:

5.7 $\quad z^2 + \tfrac{1}{2}\{a_1 \pm \sqrt{a_1^2 - 4a_2 + 4y_1}\}z + \tfrac{1}{2}\{y_1 \mp \sqrt{y_1^2 - 4a_4}\} = 0$

Suppose that all roots of 5.6 are real; then computation is simplified by using the particular real root that produces all real coefficients in the quadratic equation 5.7.

5.8
$$\begin{cases} x_1 + x_2 + x_3 + x_4 = -a_1 \\ x_1x_2 + x_2x_3 + x_3x_4 + x_4x_1 + x_1x_3 + x_2x_4 = a_2 \\ x_1x_2x_3 + x_2x_3x_4 + x_1x_2x_4 + x_1x_3x_4 = -a_3 \\ x_1x_2x_3x_4 = a_4 \end{cases}$$

where x_1, x_2, x_3, x_4 are the four roots.

6. Conversion Factors

Length

1 kilometer (km)   = 1000 meters (m)
1 meter (m)   = 100 centimeters (cm)
1 centimeter (cm) = 10^{-2} m
1 millimeter (mm) = 10^{-3} m
1 micron (μ)   = 10^{-6} m
1 millimicron (mμ) = 10^{-9} m
1 angstrom (A)   = 10^{-10} m

1 inch (in.)   = 2.540 cm
1 foot (ft)   = 30.48 cm
1 mile (mi)   = 1.609 km
1 mil   = 10^{-3} in.
1 centimeter = 0.3937 in.
1 meter   = 39.37 in.
1 kilometer = 0.6214 mile

Area

1 square meter (m^2) = 10.76 ft^2
1 square foot (ft^2)   = 929 cm^2

1 square mile (mi^2) = 640 acres
1 acre   = 43,560 ft^2

Volume

1 liter (l) = 1000 cm^3 = 1.057 quart (qt) = 61.02 in^3 = 0.03532 ft^3
1 cubic meter (m^3) = 1000 l = 35.32 ft^3
1 cubic foot (ft^3) = 7.481 U.S. gal = 0.02832 m^3 = 28.32 l
1 U.S. gallon (gal) = 231 in^3 = 3.785 l;

1 British gallon = 1.201 U.S. gallon = 277.4 in^3

Mass

1 kilogram (kg) = 2.2046 pounds (lb) = 0.6852 slug:
1 lb = 453.6 gm = 0.03108 slug
1 slug = 32.174 lb = 14.59 kg

Speed

1 km/hr = 0.2778 m/sec = 0.6214 mi/hr = 0.9113 ft/sec
1 mi/hr = 1.467 ft/sec = 1.609 km/hr = 0.4470 m/sec

Density

1 gm/cm^3 = 10^3 kg/m^3 = 62.43 lb/ft^3 = 1.940 slug/ft^3
1 lb/ft^3 = 0.01602 gm/cm^3;   1 slug/ft^3 = 0.5154 gm/cm^3

Force

1 newton (nt) = 10^5 dynes = 0.1020 kgwt = 0.2248 lbwt
1 pound weight (lbwt) = 4.448 nt = 0.4536 kgwt = 32.17 poundals
1 kilogram weight (kgwt) = 2.205 lbwt = 9.807 nt
1 U.S. short ton = 2000 lbwt:
1 long ton = 2240 lbwt;   1 metric ton = 2205 lbwt

Energy

1 joule = 1 nt m = 10^7 ergs = 0.7376 ft lbwt
  = 0.2389 cal = 9.481×10^{-4} Btu
1 ft lbwt = 1.356 joules = 0.3239 cal = 1.285×10^{-3} Btu
1 calorie (cal) = 4.186 joules = 3.087 ft lbwt = 3.968×10^{-3} Btu
1 Btu (British thermal unit) = 778 ft lbwt
  = 1055 joules = 0.293 watt hr
1 kilowatt hour (kw hr) = 3.60×10^6 joules = 860.0 kcal = 3413 Btu
1 electron volt (ev) = 1.602×10^{-19} joule

Power

1 watt = 1 joule/sec = 10^7 ergs/sec = 0.2389 cal/sec
1 horsepower (hp) = 550 ft lbwt/sec
  = 33,000 ft lbwt/min = 745.7 watts
1 kilowatt (kw) = 1.341 hp = 737.6 ft lbwt/sec = 0.9483 Btu/sec

Pressure

1 nt/m^2 = 10 dynes/cm^2 = 9.869×10^{-6} atmosphere
  = 2.089×10^{-2} lbwt/ft^2
1 lbwt/in^2 = 6895 nt/m^2 = 5.171 cm mercury
  = 27.68 in. water
1 atmosphere (atm) = 1.013×10^5 nt/m^2 = 1.013×10^6 dynes/cm^2
  = 14.70 lbwt/in^2 = 76 cm mercury = 406.8 in. water

Section II
GEOMETRY

7. Geometric Formulas

Rectangle of Length b and Width a

7.1 **Area** $= ab$

7.2 **Perimeter** $= 2a + 2b$

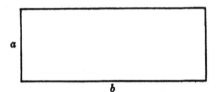

Figure 7-1

Parallelogram of Altitude h and Base b

7.3 **Area** $= bh = ab \sin \theta$

7.4 **Perimeter** $= 2a + 2b$

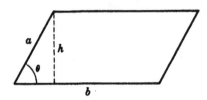

Figure 7-2

18

Triangle of Altitude *h* and Base *b*

7.5 **Area** $= \frac{1}{2}bh = \frac{1}{2}ab \sin \theta$
$$= \sqrt{s(s-a)(s-b)(s-c)}$$
where $s = \frac{1}{2}(a+b+c)$ = semiperimeter

7.6 **Perimeter** $= a + b + c$

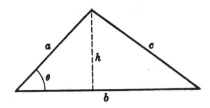

Figure 7-3

Trapezoid of Altitude *h* and Parallel Sides *a* and *b*

7.7 **Area** $= \frac{1}{2}h(a+b)$

7.8 **Perimeter** $= a + b + h\left(\dfrac{1}{\sin \theta} + \dfrac{1}{\sin \phi}\right)$
$$= a + b + h(\csc \theta + \csc \phi)$$

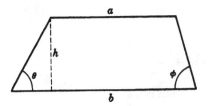

Figure 7-4

Regular Polygon of *n* Sides Each of Length *b*

7.9 **Area** $= \frac{1}{4}nb^2 \cot \dfrac{\pi}{n} = \frac{1}{4}nb^2 \dfrac{\cos(\pi/n)}{\sin(\pi/n)}$

7.10 Perimeter = nb

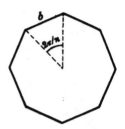

Figure 7-5

Circle of Radius r

7.11 Area = πr^2

7.12 Perimeter = $2\pi r$

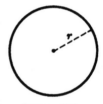

Figure 7-6

Sector of Circle of Radius r

7.13 Area = $\frac{1}{2}r^2\theta$ [θ in radians]

7.14 Arc length $s = r\theta$

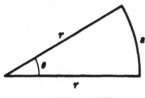

Figure 7-7

Radius of Circle Inscribed in a Triangle of Sides *a, b, c*

7.15
$$r = \frac{\sqrt{s(s-a)(s-b)(s-c)}}{s}$$

where $s = \frac{1}{2}(a + b + c)$ = semiperimeter.

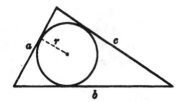

Figure 7-8

Radius of Circle Circumscribing a Triangle of Sides *a, b, c*

7.16
$$R = \frac{abc}{4\sqrt{s(s-a)(s-b)(s-c)}}$$

where $s = \frac{1}{2}(a + b + c)$ = semiperimeter.

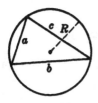

Figure 7-9

Regular Polygon of *n* Sides Inscribed in Circle of Radius *r*

7.17   Area $= \frac{1}{2}nr^2 \sin\frac{2\pi}{n} = \frac{1}{2}nr^2 \sin\frac{360°}{n}$

7.18   Perimeter $= 2nr \sin\frac{\pi}{n} = 2nr \sin\frac{180°}{n}$

Figure 7-10

Regular Polygon of *n* Sides Circumscribing a Circle of Radius *r*

7.19   Area $= nr^2 \tan\frac{\pi}{n} = nr^2 \tan\frac{180°}{n}$

7.20   Perimeter $= 2nr \tan\frac{\pi}{n} = 2nr \tan\frac{180°}{n}$

Figure 7-11

Segment of Circle of Radius *r*

7.21   **Area of shaded part** $= \frac{1}{2}r^2(\theta - \sin\theta)$

Figure 7-12

Ellipse of Semi-Major Axis *a* and Semi-Minor Axis *b*

7.22   Area = πab

7.23   Perimeter = $4a \displaystyle\int_0^{\pi/2} \sqrt{1 - k^2 \sin^2 \theta}\, d\theta$

$\qquad\qquad = 2\pi\sqrt{\tfrac{1}{2}(a^2 + b^2)}$ [approximately]

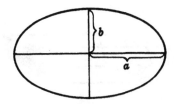

Figure 7-13

Segment of a Parabola

7.24   Area = $\tfrac{2}{3}ab$

7.25   Arc length $ABC = \tfrac{1}{2}\sqrt{b^2 + 16a^2} + \dfrac{b^2}{8a}\ln\left(\dfrac{4a + \sqrt{b^2 + 16a^2}}{b}\right)$

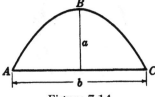

Figure 7-14

Rectangular Parallelepiped of Length *a*, Height *l*, Width *c*

7.26 **Volume** $= abc$

7.27 **Surface area** $= 2(ab + ac + bc)$

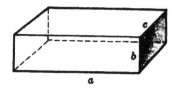

Figure 7-15

Parallelepiped of Cross-sectional Area *A* and Height *h*

7.28 **Volume** $= Ah = abc \sin \theta$

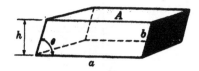

Figure 7-16

Sphere of Radius *r*

7.29 **Volume** $= \dfrac{4}{3} \pi r^3$

7.30 **Surface area** $= 4\pi r^2$

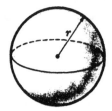

Figure 7-17

Right Circular Cylinder of Radius *r* and Height *h*

7.31 Volume = $\pi r^2 h$

7.32 Lateral surface area = $2\pi r h$

Figure 7-18

Circular Cylinder of Radius *r* and Slant Height *l*

7.33 Volume = $\pi r^2 h = \pi r^2 l \sin \theta$

7.34 Lateral surface area = $2\pi r l = \dfrac{2\pi r h}{\sin \theta} = 2\pi r h \csc \theta$

Figure 7-19

Cylinder of Cross-sectional Area *A* and Slant Height *l*

7.35 Volume = $Ah = Al \sin \theta$

7.36 Lateral surface area = $ph = pl \sin \theta$

Note that formulas 7.31 to 7.34 are special cases of formulas 7.35 and 7.36

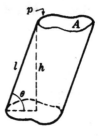

Figure 7-20

Right Circular Cone of Radius *r* and Height *h*

7.37 Volume $= \frac{1}{3}\pi r^2 h$

7.38 Lateral surface area $= \pi r \sqrt{r^2 + h^2} = \pi r l$

Figure 7-21

Pyramid of Base Area *A* and Height *h*

7.39 Volume $= \frac{1}{3}Ah$

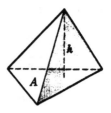

Figure 7-22

Spherical Cap of Radius *r* and Height *h*

7.40 **Volume (shaded in figure)** $= \frac{1}{3}\pi h^2(3r - h)$

7.41 **Surface area** $= 2\pi rh$

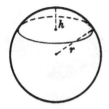

Figure 7-23

Frustum of Right Circular Cone of Radii *a*, *b* and Height *h*

7.42 **Volume** $= \frac{1}{3}\pi h(a^2 + ab + b^2)$

7.43 **Lateral surface area** $= \pi(a + b)\sqrt{h^2 + (b - a)^2}$
$$= \pi(a + b)l$$

Figure 7-24

Spherical Triangle of Angles *A*, *B*, *C* on Sphere of Radius *r*

7.44 **Area of triangle** $ABC = (A + B + C - \pi)r^2$

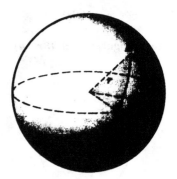

Figure 7-25

Torus of Inner Radius *a* and Outer Radius *b*

7.45   Volume $= \frac{1}{4}\pi^2(a+b)(b-a)^2$

7.46   Surface area $= \pi^2(b^2 - a^2)$

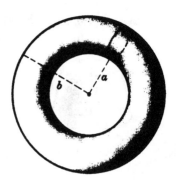

Figure 7-26

Ellipsoid of Semi-Axes *a, b, c*

7.47   Volume $= \frac{4}{3}\pi abc$

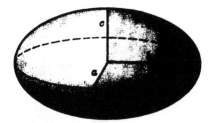

Figure 7-27

Paraboloid of Revolution

7.48 **Volume** $= \frac{1}{2}\pi b^2 a$

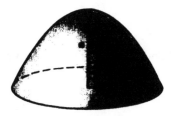

Figure 7-28

8. Formulas from Plane Analytic Geometry

Distance d Between Two Points $P_1(x_1, y_1)$ and $P_2(x_2, y_2)$

8.1 $\quad d = \sqrt{(x_2 - x_1)^2 + (y_2 - y_1)^2}$

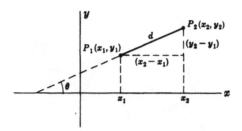

Figure 8-1

Slope m of Line Joining Two Points $P_1(x_1,y_1)$ and $P_2(x_2,y_2)$

8.2 $m = \dfrac{y_2 - y_1}{x_2 - x_1} = \tan \theta$

Equation of Line Joining Two Points $P_1(x_1,y_1)$ and $P_2(x_2,y_2)$

8.3 $\dfrac{y - y_1}{x - x_1} = \dfrac{y_2 - y_1}{x_2 - x_1} = m$ or $y - y_1 = m(x - x_1)$

8.4 $y = mx + b$

where $b = y_1 - mx_1 = \dfrac{x_2 y_1 - x_1 y_2}{x_2 - x_1}$ is the *intercept* on the y axis, i.e. the y *intercept*.

Equation of Line in Terms of x Intercept $a \neq 0$ and y Intercept $b \neq 0$

8.5 $\dfrac{x}{a} + \dfrac{y}{b} = 1$

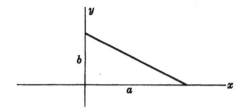

Figure 8-2

Normal Form for Equation of Line

8.6 $x \cos \alpha + y \sin \alpha = p$

where p = perpendicular distance from origin O to line and α = angle of inclination of perpendicular with positive x axis

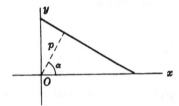

Figure 8-3

General Equation of Line

8.7 $Ax + By + C = 0$

Distance from Point (x_1, y_1) to Line $Ax + By + C = 0$

8.8 $\dfrac{Ax_1 + By_1 + C}{\pm \sqrt{A^2 + B^2}}$

where the sign is chosen so that the distance is nonnegative.

Angle ψ Between Two Lines Having Slopes m_1 and m_2

8.9 $\tan \psi = \dfrac{m_2 - m_1}{1 + m_1 m_2}$

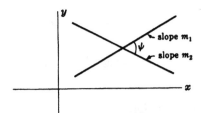

Figure 8-4

Lines are parallel or coincident if and only if $m_1 = m_2$.
Lines are perpendicular if and only if $m_2 = -1/m_1$.

Area of Triangle With Vertices at (x_1, y_1), (x_2, y_2), (x_3, y_3)

8.10 $\text{Area} = \pm\dfrac{1}{2} \begin{vmatrix} x_1 & y_1 & 1 \\ x_2 & y_2 & 1 \\ x_3 & y_3 & 1 \end{vmatrix}$

$= \pm\dfrac{1}{2}(x_1 y_2 + y_1 x_3 + y_3 x_2 - y_2 x_3 - y_1 x_2 - x_1 y_3)$

where the sign is chosen so that the area is nonnegative.
If the area is zero, the points all lie on a line.

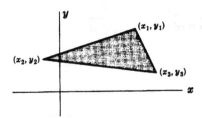

Figure 8-5

Transformation of Coordinates Involving Pure Translation

8.11
$$\begin{cases} x = x' + x_0 \\ y = y' + y_0 \end{cases} \quad \text{or} \quad \begin{cases} x' = x - x_0 \\ y' = y - y_0 \end{cases}$$

where (x, y) are old coordinates [i.e. coordinates relative to xy system], (x', y') are new coordinates [relative to $x'y'$ system], and (x_0, y_0) are the coordinates of the new origin O' relative to the old xy coordinate system.

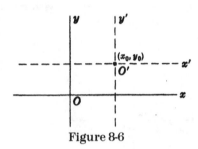

Figure 8-6

Transformation of Coordinates Involving Pure Rotation

8.12
$$\begin{cases} x = x' \cos \alpha - y' \sin \alpha \\ y = x' \sin \alpha + y' \cos \alpha \end{cases} \quad \text{or} \quad \begin{cases} x' = x \cos \alpha + y \sin \alpha \\ y' = y \cos \alpha - x \sin \alpha \end{cases}$$

where the origins of the old $[xy]$ and new $[x'y']$ coordinate systems are the same but the x' axis makes an angle α with the positive x axis.

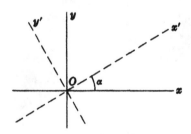

Figure 8-7

Transformation of Coordinates Involving Translation and Rotation

8.13
$$\begin{cases} x = x' \cos \alpha - y' \sin \alpha + x_0 \\ y = x' \sin \alpha + y' \cos \alpha + y_0 \end{cases}$$

or
$$\begin{cases} x' = (x - x_0) \cos \alpha + (y - y_0) \sin \alpha \\ y' = (y - y_0) \cos \alpha - (x - x_0) \sin \alpha \end{cases}$$

where the new origin O' of $x'y'$ coordinate system has coordinates (x_0, y_0) relative to the old xy coordinate system and the x' axis makes an angle α with the positive x axis.

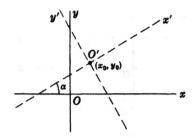

Figure 8-8

Polar Coordinates (r, θ)

A point P can be located by rectangular coordinates (x, y) or polar coordinates (r, θ). The transformation between these coordinates is as follows:

8.14
$$\begin{cases} x = r \cos \theta \\ y = r \sin \theta \end{cases} \quad \text{or} \quad \begin{cases} r = \sqrt{x^2 + y^2} \\ \theta = \tan^{-1}(y/x) \end{cases}$$

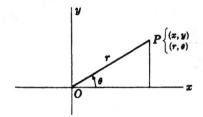

Figure 8-9

Equation of Circle of Radius R, Center at (x_o, y_o)

8.15 $(x - x_0)^2 + (y - y_0)^2 = R^2$

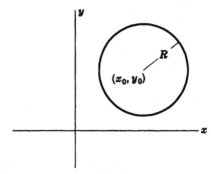

Figure 8-10

Equation of Circle of Radius R Passing Through Origin

8.16 $r = 2R \cos(\theta - \alpha)$

where (r, θ) are polar coordinates of any point on the circle and (R, α) are polar coordinates of the center of the circle.

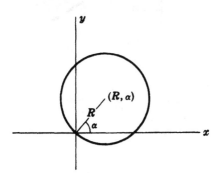

Figure 8-11

Conics [Ellipse, Parabola, or Hyperbola]

If a point P moves so that its distance from a fixed point [called the *focus*] divided by its distance from a fixed line [called the *directrix*] is a constant ϵ [called the *eccentricity*], then the curve described by P is called a *conic* [so-called because such curves can be obtained by intersecting a plane and a cone at different angles].

If the focus is chosen at origin O the equation of a conic in polar coordinates (r, θ) is, if $OQ = p$ and $LM = D$ [see Fig. 8-12]

8.17 $\qquad r = \dfrac{p}{1 - \epsilon \cos \theta} = \dfrac{\epsilon D}{1 - \epsilon \cos \theta}$

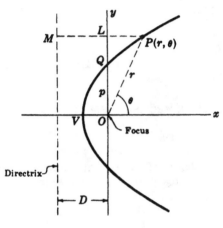

Figure 8-12

The conic is

    (1) an ellipse if $\epsilon < 1$
    (2) a parabola if $\epsilon = 1$
    (3) a hyperbola if $\epsilon > 1$

Ellipse With Center $C(x_0, y_0)$ and Major Axis Parallel to x Axis

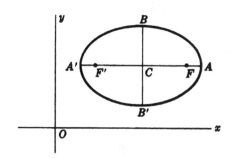

Figure 8-13

8.18 Length of major axis $A'A = 2a$

8.19 Length of minor axis $B'B = 2b$

8.20 Distance from center C to focus F or F' is
$$c = \sqrt{a^2 - b^2}$$

8.21 Eccentricity $= \epsilon = \dfrac{c}{a} = \dfrac{\sqrt{a^2 - b^2}}{a}$

8.22 Equation in rectangular coordinates:
$$\frac{(x - x_0)^2}{a^2} + \frac{(y - y_0)^2}{b^2} = 1$$

8.23 Equation in polar coordinates if C is at O: $r^2 = \dfrac{a^2 b^2}{a^2 \sin^2 \theta + b}$

8.24

Equation in polar coordinates if C is at O: $r^2 = \dfrac{a^2 b^2}{a^2 \sin^2 \theta + b^2 \cos^2 \theta}$

8.25 If P is any point on the ellipse, $PF + PF' = 2a$

If the major axis is parallel to the y-axis, interchange x and y in the above or replace θ by $90° - \theta$.

Parabola With Axis Parallel to x Axis

If vertex is at $A(x_0, y_0)$ and the distance from A to focus F is $a > 0$, the equation of the parabola is

8.26 $(y - y_0)^2 = 4a(x - x_0)$

if parabola opens to right [Fig. 8-14]

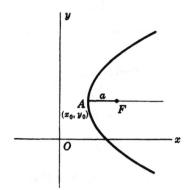

Figure 8-14

8.27 $(y - y_0)^2 = -4a(x - x_0)$

if parabola opens to left [Fig. 8-15]

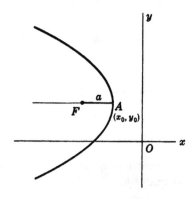

Figure 8-15

If the focus is at the origin [Fig. 8-16], the equation in polar coordinates is

8.28 $r = \dfrac{2a}{1 - \cos \theta}$

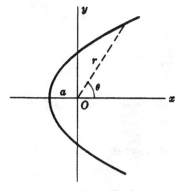

Figure 8-16

If the axis is parallel to the y axis, interchange x and y in the above or replace θ by $90° - \theta$.

Hyperbola With Center $C(x_0, y_0)$ and Major Axis Parallel to x Axis

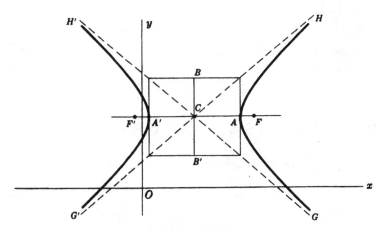

Figure 8-17

8.29 **Length of major axis $A'A = 2a$**

8.30 **Length of minor axis $B'B = 2b$**

8.31 **Distance from center C to focus F or $F' = c = \sqrt{a^2 + b^2}$**

8.32 **Eccentricity $\epsilon = \dfrac{c}{a} = \dfrac{\sqrt{a^2 + b^2}}{a}$**

8.33 **Equation in rectangular coordinates:** $\dfrac{(x - x_0)^2}{a^2} - \dfrac{(y - y_0)^2}{b^2} = 1$

8.34 **Slopes of asymptotes $G'H$ and $GH' = \pm\dfrac{b}{a}$**

8.35 **Equation in polar coordinates if C is at O:** $r^2 = \dfrac{a^2 b^2}{b^2 \cos^2 \theta - a^2 \sin^2 \theta}$

8.36 **Equation in polar coordinates if C is on x axis and F' is at O:**
$$r = \frac{a(\epsilon^2 - 1)}{1 - \epsilon \cos \theta}$$

8.37 **If P is any point on the hyperbola, $PF - PF' = \pm 2a$ [depending on branch]**

If the major axis is parallel to the y axis, interchange x and y in the above or replace θ by $90° - \theta$.

9. Formulas from Solid Analytic Geometry

Distance d Between Two Points $P_1(x_1, y_1, z_1)$ and $P_2(x_2, y_2, z_2)$

9.1 $d = \sqrt{(x_2 - x_1)^2 + (y_2 - y_1)^2 + (z_2 - z_1)^2}$

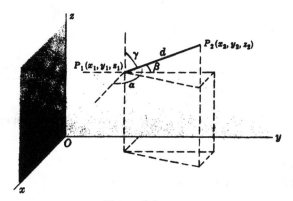

Figure 9-1

Direction Cosines of Line Joining Two Points $P_1(x_1,y_1,z_1)$ and $P_2(x_2,y_2,z_2)$

$$9.2 \quad l = \cos \alpha = \frac{x_2 - x_1}{d}, \quad m = \cos \beta = \frac{y_2 - y_1}{d}, \quad n = \cos \gamma = \frac{z_2 - z_1}{d}$$

where α, β, γ are the angles which line $P_1 P_2$ makes with the positive x, y, z axes respectively, and d is given by 9.1 [see Fig. 9-1].

Relationship Between Direction Cosines

$$9.3 \quad \cos^2 \alpha + \cos^2 \beta + \cos^2 \gamma = 1 \quad \text{or} \quad l^2 + m^2 + n^2 = 1$$

Direction Numbers

Numbers L, M, N which are proportional to the direction cosines l, m, n are called *direction numbers*. The relationship between them is given by

$$9.4 \quad l = \frac{L}{\sqrt{L^2 + M^2 + N^2}}, \quad m = \frac{M}{\sqrt{L^2 + M^2 + N^2}}, \quad n = \frac{N}{\sqrt{L^2 + M^2 + N^2}}$$

Equations of Line Joining $P_1(x_1, y_1, z_1)$ and $P_2(x_2, y_2, z_2)$ in Standard Form

9.5 $\qquad \dfrac{x - x_1}{x_2 - x_1} = \dfrac{y - y_1}{y_2 - y_1} = \dfrac{z - z_1}{z_2 - z_1}$ or $\dfrac{x - x_1}{l} = \dfrac{y - y_1}{m} = \dfrac{z - z_1}{n}$

These are also valid if l, m, n are replaced by L, M, N respectively.

Equations of Line Joining $P_1(x_1, y_1, z_1)$ and $P_2(x_2, y_2, z_2)$ in Parametric Form

9.6 $\qquad x = x_1 + lt, \quad y = y_1 + mt, \quad z = z_1 + nt$

These are also valid if l, m, n are replaced by L, M, N respectively.

Angle ϕ Between Two Lines with Direction Cosines l_1, m_1, n_1 and l_2, m_2, n_2

9.7 $\qquad \cos \phi = l_1 l_2 + m_1 m_2 + n_1 n_2$

General Equation of a Plane

9.8 $\qquad Ax + By + Cz + D = 0$

[A, B, C, D are constants]

Equation of Plane Passing Through Points (x_1, y_1, z_1), (x_2, y_2, z_2), (x_3, y_3, z_3)

9.9 $\qquad \begin{vmatrix} x - x_1 & y - y_1 & z - z_1 \\ x_2 - x_1 & y_2 - y_1 & z_2 - z_1 \\ x_3 - x_1 & y_3 - y_1 & z_3 - z_1 \end{vmatrix} = 0$

or

9.10

$$\begin{vmatrix} y_2 - y_1 & z_2 - z_1 \\ y_3 - y_1 & z_3 - z_1 \end{vmatrix} (x - x_1) + \begin{vmatrix} z_2 - z_1 & x_2 - x_1 \\ z_3 - z_1 & x_3 - x_1 \end{vmatrix} (y - y_1) + \begin{vmatrix} x_2 - x_1 & y_2 - y_1 \\ x_3 - x_1 & y_3 - y_1 \end{vmatrix} (z - z_1) = 0$$

Equation of Plane in Intercept Form

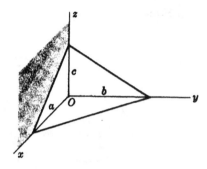

Figure 9-2

9.11 $\dfrac{x}{a} + \dfrac{y}{b} + \dfrac{z}{c} = 1$

where a, b, c are the intercepts on the x, y, z axes respectively.

Equations of Line Through Point (x_o, y_o, z_o) and Perpendicular to Plane $Ax + By + Cz + D = 0$

9.12 $\dfrac{x - x_0}{A} = \dfrac{y - y_0}{B} = \dfrac{z - z_0}{C}$ or

$x = x_0 + At, \quad y = y_0 + Bt, \quad z = z_0 + Ct$

The direction numbers for a line perpendicular to the plane $Ax + By + Cz + D = 0$ are A, B, C.

Distance From Point (x_0, y_0, z_0) to Plane $Ax + By + Cz + D = 0$

9.13 $\dfrac{Ax_0 + By_0 + Cz_0 + D}{\pm \sqrt{A^2 + B^2 + C^2}}$

where the sign is chosen so that the distance is nonnegative.

Normal Form for Equation of Plane

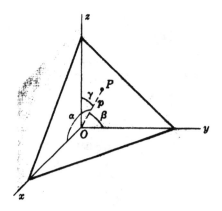

Figure 9-3

9.14 $\quad x \cos \alpha + y \cos \beta + z \cos \gamma = p$

where p = perpendicular distance from O to plane at P and α, β, γ are angles between OP and positive x, y, z axes.

Transformation of Coordinates Involving Pure Translation

9.15 $\quad \begin{cases} x = x' + x_0 \\ y = y' + y_0 \\ z = z' + z_0 \end{cases}$ or $\begin{cases} x' = x - x_0 \\ y' = y - y_0 \\ z' = z - z_0 \end{cases}$

where (x, y, z) are old coordinates [i.e. coordinates relative to xyz system], (x', y', z') are new coordinates [relative to $x'y'z'$ system] and (x_0, y_0, z_0) are the coordinates of the new origin O' relative to the old xyz coordinate system.

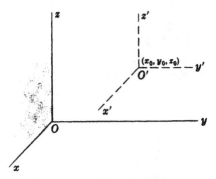

Figure 9-4

Transformation of Coordinates Involving Pure Rotation

9.16

$$\begin{cases} x = l_1 x' + l_2 y' + l_3 z' \\ y = m_1 x' + m_2 y' + m_3 z' \\ z = n_1 x' + n_2 y' + n_3 z' \end{cases}$$

or

$$\begin{cases} x' = l_1 x + m_1 y + n_1 z \\ y' = l_2 x + m_2 y + n_2 z \\ z' = l_3 x + m_3 y + n_3 z \end{cases}$$

where the origins of the xyz and $x'\,y'\,z'$ systems are the same and l_1, m_1, n_1; l_2, m_2, n_2; l_3, m_3, n_3 are the direction cosines of the x', y', z' axes relative to the x, y, z axes respectively.

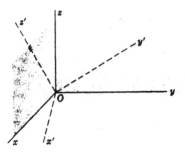

Figure 9-5

Transformation of Coordinates Involving Translation and Rotation

9.17
$$\begin{cases} x = l_1 x' + l_2 y' + l_3 z' + x_0 \\ y = m_1 x' + m_2 y' + m_3 z' + y_0 \\ z = n_1 x' + n_2 y' + n_3 z' + z_0 \end{cases}$$

or
$$\begin{cases} x' = l_1(x - x_0) + m_1(y - y_0) + n_1(z - z_0) \\ y' = l_2(x - x_0) + m_2(y - y_0) + n_2(z - z_0) \\ z' = l_3(x - x_0) + m_3(y - y_0) + n_3(z - z_0) \end{cases}$$

where the origin O' of the $x'\,y'\,z'$ system has coordinates (x_0, y_0, z_0) relative to the xyz system and $l_1,\, m_1,\, n_1;\, l_2,\, m_2,\, n_2;\, l_3,\, m_3,\, n_3$ are the direction cosines of the $x',\, y',\, z'$ axes relative to the $x,\, y,\, z$ axes respectively.

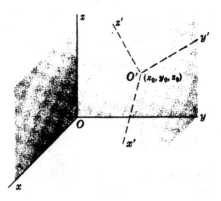

Figure 9-6

Cylindrical Coordinates (r, θ, z)

A point P can be located by cylindrical coordinates (r, θ, z) [see Fig. 9-7] as well as rectangular coordinates (x, y, z).

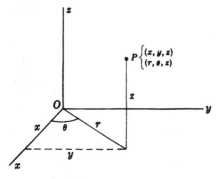

Figure 9-7

The transformation between these coordinates is:

9.18
$$\begin{cases} x = r\cos\theta \\ y = r\sin\theta \\ z = z \end{cases} \quad \text{or} \quad \begin{cases} r = \sqrt{x^2 + y^2} \\ \theta = \tan^{-1}(y/x) \\ z = z \end{cases}$$

Spherical Coordinates (r, θ, ϕ)

A point P can be located by spherical coordinates (r, θ, ϕ) [see Fig. 9-8] as well as rectangular coordinates (x, y, z).

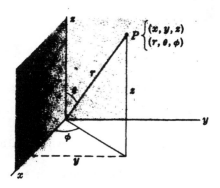

Figure 9-8

The transformation between these coordinates is:

9.19
$$\begin{cases} x = r\sin\theta\cos\phi \\ y = r\sin\theta\sin\phi \\ z = r\cos\theta \end{cases}$$

or
$$\begin{cases} r = \sqrt{x^2 + y^2 + z^2} \\ \phi = \tan^{-1}(y/x) \\ \theta = \cos^{-1}(z/\sqrt{x^2 + y^2 + z^2}) \end{cases}$$

Equation of Sphere in Rectangular Coordinates

9.20 $(x - x_0)^2 + (y - y_0)^2 + (z - z_0)^2 = R^2$

where the sphere has center (x_o, y_o, z_o) and radius R.

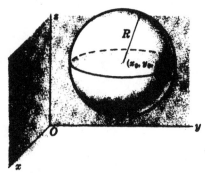

Figure 9-9

Equation of Sphere in Cylindrical Coordinates

9.21 $r^2 - 2r_0 r\cos(\theta - \theta_0) + r_0^2 + (z - z_0)^2 = R^2$

where the sphere has center (r_o, θ_o, z_o) in cylindrical coordinates and radius R.

If the center is at the origin, the equation is:

9.22 $r^2 + z^2 = R^2$

Equation of Sphere in Spherical Coordinates

9.23 $r^2 + r_0^2 - 2r_0 r \sin\theta \sin\theta_0 \cos(\phi - \phi_0) = R^2$

where the sphere has center (r_0, θ_0, ϕ_0) in spherical coordinates and radius R.

If the center is at the origin, the equation is:

9.24 $r = R$

Equation of Ellipsoid with Center (x_0, y_0, z_0) and Semi-axes a, b, c

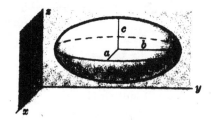

Figure 9-10

9.25 $\dfrac{(x - x_0)^2}{a^2} + \dfrac{(y - y_0)^2}{b^2} + \dfrac{(z - z_0)^2}{c^2} = 1$

Elliptic Cylinder with Axis as z Axis

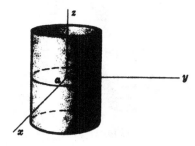

Figure 9-11

9.26 $\quad \dfrac{x^2}{a^2} + \dfrac{y^2}{b^2} = 1$

where a, b are semi-axes of elliptic cross section.

If $b = a$, it becomes a circular cylinder of radius a.

Elliptic Cone with Axis as z Axis

Figure 9-12

9.27 $\quad \dfrac{x^2}{a^2} + \dfrac{y^2}{b^2} = \dfrac{z^2}{c^2}$

Hyperboloid of One Sheet

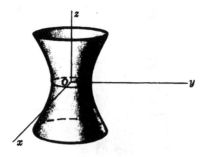

Figure 9-13

9.28 $\dfrac{x^2}{a^2} + \dfrac{y^2}{b^2} - \dfrac{z^2}{c^2} = 1$

Hyperboloid of Two Sheets

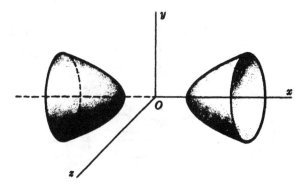

Figure 9-14

9.29 $\dfrac{x^2}{a^2} - \dfrac{y^2}{b^2} - \dfrac{z^2}{c^2} = 1$

Note orientation of axes in Fig. 9-14.

Elliptic Paraboloid

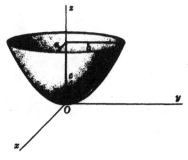

Figure 9-15

9.30 $\quad \dfrac{x^2}{a^2} + \dfrac{y^2}{b^2} = \dfrac{z}{c}$

Hyperbolic Paraboloid

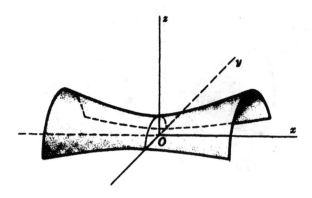

Figure 9-16

9.31 $\quad \dfrac{x^2}{a^2} - \dfrac{y^2}{b^2} = \dfrac{z}{c}$

Note orientation of axes in Fig. 9-16.

10. Trigonometric Functions

Definition of Trigonometric Functions for a Right Triangle

Triangle ABC has a right angle (90°) at C and sides of length a, b, c.

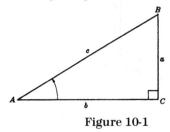

Figure 10-1

The trigonometric functions of angle A are defined as follows:

10.1 $\quad$ **sine** of A = $\sin A = \dfrac{a}{c} = \dfrac{\text{opposite}}{\text{hypotenuse}}$

10.2 $\quad$ **cosine** of A = $\cos A = \dfrac{b}{c} = \dfrac{\text{adjacent}}{\text{hypotenuse}}$

10.3 $\quad$ **tangent** of A = $\tan A = \dfrac{a}{b} = \dfrac{\text{opposite}}{\text{adjacent}}$

10.4 $\quad$ **cotangent** of A = $\cot A = \dfrac{b}{a} = \dfrac{\text{adjacent}}{\text{opposite}}$

10.5 $\quad$ **secant** of A = $\sec A = \dfrac{c}{b} = \dfrac{\text{hypotenuse}}{\text{adjacent}}$

10.6   *cosecant* of $A = \csc A = \dfrac{c}{a} = \dfrac{\text{hypotenuse}}{\text{opposite}}$

Extensions to Angles Which May Be Greater than 90°

Consider an xy coordinate system [see Figs. 10-2 and 10-3 below]. A point P in the xy plane has coordinates (x, y) where x is considered as positive along OX and negative along OX' while y is considered as positive along OY and negative along OY'. The distance from origin O to point P is positive and is related through the Pythagorean Theorem [$r^2 = x^2 + y^2$]. The angle A described *counterclockwise* from OX is considered *positive*. If it is described *clockwise* from OX, it is considered *negative*.

The various quadrants are denoted by I, II, III, and IV called the first, second, third, and fourth quadrants, respectively. In Fig. 10-2, for example, angle A is in the second quadrant while in Fig. 10-3, angle A is in the third quadrant.

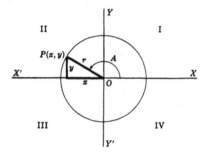

Figure 10-2

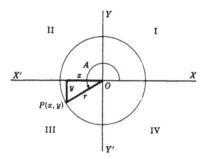

Figure 10-3

For an angle A in any quadrant the trigonometric functions of A are defined as follows:

10.7 $\quad \sin A = y/r$

10.8 $\quad \cos A = x/r$

10.9 $\quad \tan A = y/x$

10.10 $\quad \cot A = x/y$

10.11 $\quad \sec A = r/x$

10.12 $\quad \csc A = r/y$

Relationship Between Degrees and Radians

A *radian* is that angle θ subtended at center O of a circle by an arc MN equal to the radius r, as shown in Fig. 10-4.

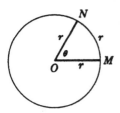

Figure 10-4

Since 2π radians = 360°, we have

10.13 1 radian = $180°/\pi$ = 57.29577 95130 8232 ...°

10.14 1° = $\pi/180$ radians = 0.01745 32925 19943 29576 92 ... radians

Relationships Among Trigonometric Functions

10.15 $\quad \tan A = \dfrac{\sin A}{\cos A}$

10.16 $\quad \cot A = \dfrac{1}{\tan A} = \dfrac{\cos A}{\sin A}$

10.17 $\quad \sec A = \dfrac{1}{\cos A}$

10.18 $\quad \csc A = \dfrac{1}{\sin A}$

10.19 $\quad \sin^2 A + \cos^2 A = 1$

10.20 $\quad \sec^2 A - \tan^2 A = 1$

10.21 $\quad \csc^2 A - \cot^2 A = 1$

Exact Values for Trigonometric Functions of Various Angles

Angle A in degrees	Angle A in radians	sin A	cos A	tan A	cot A	sec A	csc A
0°	0	0	1	0	∞	1	∞
15°	$\pi/12$	$\frac{1}{4}(\sqrt{6}-\sqrt{2})$	$\frac{1}{4}(\sqrt{6}+\sqrt{2})$	$2-\sqrt{3}$	$2+\sqrt{3}$	$\sqrt{6}-\sqrt{2}$	$\sqrt{6}+\sqrt{2}$
30°	$\pi/6$	$\frac{1}{2}$	$\frac{1}{2}\sqrt{3}$	$\frac{1}{3}\sqrt{3}$	$\sqrt{3}$	$\frac{2}{3}\sqrt{3}$	2
45°	$\pi/4$	$\frac{1}{2}\sqrt{2}$	$\frac{1}{2}\sqrt{2}$	1	1	$\sqrt{2}$	$\sqrt{2}$
60°	$\pi/3$	$\frac{1}{2}\sqrt{3}$	$\frac{1}{2}$	$\sqrt{3}$	$\frac{1}{3}\sqrt{3}$	2	$\frac{2}{3}\sqrt{3}$
75°	$5\pi/12$	$\frac{1}{4}(\sqrt{6}+\sqrt{2})$	$\frac{1}{4}(\sqrt{6}-\sqrt{2})$	$2+\sqrt{3}$	$2-\sqrt{3}$	$\sqrt{6}+\sqrt{2}$	$\sqrt{6}-\sqrt{2}$
90°	$\pi/2$	1	0	$\pm\infty$	0	$\pm\infty$	1
105°	$7\pi/12$	$\frac{1}{4}(\sqrt{6}+\sqrt{2})$	$-\frac{1}{4}(\sqrt{6}-\sqrt{2})$	$-(2+\sqrt{3})$	$-(2-\sqrt{3})$	$-(\sqrt{6}+\sqrt{2})$	$\sqrt{6}-\sqrt{2}$
120°	$2\pi/3$	$\frac{1}{2}\sqrt{3}$	$-\frac{1}{2}$	$-\sqrt{3}$	$-\frac{1}{3}\sqrt{3}$	-2	$\frac{2}{3}\sqrt{3}$
135°	$3\pi/4$	$\frac{1}{2}\sqrt{2}$	$-\frac{1}{2}\sqrt{2}$	-1	-1	$-\sqrt{2}$	$\sqrt{2}$
150°	$5\pi/6$	$\frac{1}{2}$	$-\frac{1}{2}\sqrt{3}$	$-\frac{1}{3}\sqrt{3}$	$-\sqrt{3}$	$-\frac{2}{3}\sqrt{3}$	2

Angle	Radians	sin	cos	tan	cot	sec	csc
165°	$11\pi/12$	$\frac{1}{4}(\sqrt{6}-\sqrt{2})$	$-\frac{1}{4}(\sqrt{6}+\sqrt{2})$	$-(2-\sqrt{3})$	$-(2+\sqrt{3})$	$-(\sqrt{6}-\sqrt{2})$	$\sqrt{6}+\sqrt{2}$
180°	π	0	-1	0	$\pm\infty$	-1	$\pm\infty$
195°	$13\pi/12$	$-\frac{1}{4}(\sqrt{6}-\sqrt{2})$	$-\frac{1}{4}(\sqrt{6}+\sqrt{2})$	$2-\sqrt{3}$	$2+\sqrt{3}$	$-(\sqrt{6}-\sqrt{2})$	$-(\sqrt{6}+\sqrt{2})$
210°	$7\pi/6$	$-\frac{1}{2}$	$-\frac{1}{2}\sqrt{3}$	$\frac{1}{3}\sqrt{3}$	$\sqrt{3}$	$-\frac{2}{3}\sqrt{3}$	-2
225°	$5\pi/4$	$-\frac{1}{2}\sqrt{2}$	$-\frac{1}{2}\sqrt{2}$	1	1	$-\sqrt{2}$	$-\sqrt{2}$
240°	$4\pi/3$	$-\frac{1}{2}\sqrt{3}$	$-\frac{1}{2}$	$\sqrt{3}$	$\frac{1}{3}\sqrt{3}$	-2	$-\frac{2}{3}\sqrt{3}$
255°	$17\pi/12$	$-\frac{1}{4}(\sqrt{6}+\sqrt{2})$	$-\frac{1}{4}(\sqrt{6}-\sqrt{2})$	$2+\sqrt{3}$	$2-\sqrt{3}$	$-(\sqrt{6}+\sqrt{2})$	$-(\sqrt{6}-\sqrt{2})$
270°	$3\pi/2$	-1	0	$\pm\infty$	0	$\pm\infty$	-1
285°	$19\pi/12$	$-\frac{1}{4}(\sqrt{6}+\sqrt{2})$	$\frac{1}{4}(\sqrt{6}-\sqrt{2})$	$-(2+\sqrt{3})$	$-(2-\sqrt{3})$	$\sqrt{6}+\sqrt{2}$	$-(\sqrt{6}-\sqrt{2})$
300°	$5\pi/3$	$-\frac{1}{2}\sqrt{3}$	$\frac{1}{2}$	$-\sqrt{3}$	$-\frac{1}{3}\sqrt{3}$	2	$-\frac{2}{3}\sqrt{3}$
315°	$7\pi/4$	$-\frac{1}{2}\sqrt{2}$	$\frac{1}{2}\sqrt{2}$	-1	-1	$\sqrt{2}$	$-\sqrt{2}$
330°	$11\pi/6$	$-\frac{1}{2}$	$\frac{1}{2}\sqrt{3}$	$-\frac{1}{3}\sqrt{3}$	$-\sqrt{3}$	$\frac{2}{3}\sqrt{3}$	-2
345°	$23\pi/12$	$-\frac{1}{4}(\sqrt{6}-\sqrt{2})$	$\frac{1}{4}(\sqrt{6}+\sqrt{2})$	$-(2-\sqrt{3})$	$-(2+\sqrt{3})$	$\sqrt{6}-\sqrt{2}$	$-(\sqrt{6}+\sqrt{2})$
360°	2π	0	1	0	$\pm\infty$	1	$\pm\infty$

Graphs of Trigonometric Functions

In each graph, x is in radians.

10.22 $y = \sin x$

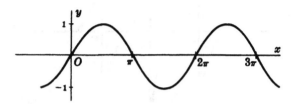

Figure 10-5

10.23 $y = \cos x$

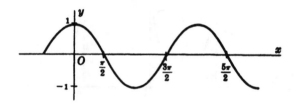

Figure 10-6

10.24 $y = \tan x$

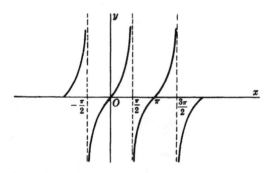

Figure 10-7

10.25 $y = \cot x$

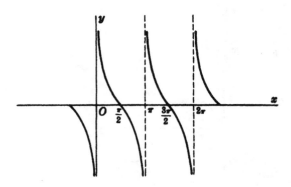

Figure 10-8

10.26 $y = \sec x$

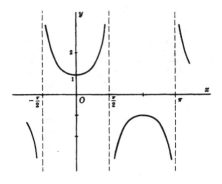

Figure 10-9

10.27 $y = \csc x$

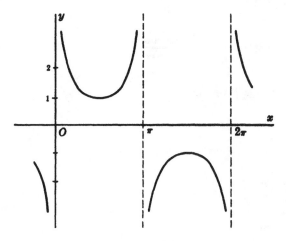

Figure 10-10

Functions of Negative Angles

10.28 $\sin(-A) = -\sin A$

10.29 $\cos(-A) = \cos A$

10.30 $\tan(-A) = -\tan A$

10.31 $\csc(-A) = -\csc A$

10.32 $\sec(-A) = \sec A$

10.33 $\cot(-A) = -\cot A$

Addition Formulas

10.34 $\sin(A \pm B) = \sin A \cos B \pm \cos A \sin B$

10.35 $\cos(A \pm B) = \cos A \cos B \mp \sin A \sin B$

10.36 $\quad \tan(A \pm B) = \dfrac{\tan A \pm \tan B}{1 \mp \tan A \tan B}$

10.37 $\quad \cot(A \pm B) = \dfrac{\cot A \cot B \mp 1}{\cot B \pm \cot A}$

Double Angle Formulas

10.38 $\quad \sin 2A = 2 \sin A \cos A$

10.39 $\quad \cos 2A = \cos^2 A - \sin^2 A = 1 - 2\sin^2 A = 2\cos^2 A - 1$

10.40 $\quad \tan 2A = \dfrac{2\tan A}{1 - \tan^2 A}$

Half Angle Formulas

10.41 $\quad \sin \dfrac{A}{2} = \pm \sqrt{\dfrac{1 - \cos A}{2}} \quad \begin{bmatrix} + \text{ if } A/2 \text{ is in quadrant I or II} \\ - \text{ if } A/2 \text{ is in quadrant III or IV} \end{bmatrix}$

10.42 $\quad \cos \dfrac{A}{2} = \pm \sqrt{\dfrac{1 + \cos A}{2}} \quad \begin{bmatrix} + \text{ if } A/2 \text{ is in quadrant I or IV} \\ - \text{ if } A/2 \text{ is in quadrant II or III} \end{bmatrix}$

10.43 $\quad \tan \dfrac{A}{2} = \pm \sqrt{\dfrac{1 - \cos A}{1 + \cos A}} \quad \begin{bmatrix} + \text{ if } A/2 \text{ is in quadrant I or III} \\ - \text{ if } A/2 \text{ is in quadrant II or IV} \end{bmatrix}$

$$= \dfrac{\sin A}{1 + \cos A} = \dfrac{1 - \cos A}{\sin A} = \csc A - \cot A$$

Powers of Trigonometric Functions

10.44 $\quad \sin^2 A = \tfrac{1}{2} - \tfrac{1}{2}\cos 2A$

10.45 $\quad \cos^2 A = \tfrac{1}{2} + \tfrac{1}{2}\cos 2A$

10.46 $\quad \sin^3 A = \tfrac{3}{4}\sin A - \tfrac{1}{4}\sin 3A$

10.47 $\quad \cos^3 A = \tfrac{3}{4}\cos A + \tfrac{1}{4}\cos 3A$

10.48 $\quad \sin^4 A = \tfrac{3}{8} - \tfrac{1}{2}\cos 2A + \tfrac{1}{8}\cos 4A$

10.49 $\cos^4 A = \frac{3}{8} + \frac{1}{2}\cos 2A + \frac{1}{8}\cos 4A$

10.50 $\sin^5 A = \frac{5}{8}\sin A - \frac{5}{16}\sin 3A + \frac{1}{16}\sin 5A$

10.51 $\cos^5 A = \frac{5}{8}\cos A + \frac{5}{16}\cos 3A + \frac{1}{16}\cos 5A$

Sum, Difference, and Product of Trigonometric Functions

10.52 $\sin A + \sin B = 2\sin\frac{1}{2}(A + B)\cos\frac{1}{2}(A - B)$

10.53 $\sin A - \sin B = 2\cos\frac{1}{2}(A + B)\sin\frac{1}{2}(A - B)$

10.54 $\cos A + \cos B = 2\cos\frac{1}{2}(A + B)\cos\frac{1}{2}(A - B)$

10.55 $\cos A - \cos B = 2\sin\frac{1}{2}(A + B)\sin\frac{1}{2}(B - A)$

10.56 $\sin A \sin B = \frac{1}{2}\{\cos(A - B) - \cos(A + B)\}$

10.57 $\cos A \cos B = \frac{1}{2}\{\cos(A - B) + \cos(A + B)\}$

10.58 $\sin A \cos B = \frac{1}{2}\{\sin(A - B) + \sin(A + B)\}$

Inverse Trigonometric Functions

If $x = \sin y$, then $y = \sin^{-1} x$, i.e., *the angle whose sine is x or inverse sine of x*, is a many-valued function of x which is a collection of single-valued functions called *branches*. Similarly, the other inverse trigonometric functions are multiple-valued.

For many purposes, a particular branch is required. This is called the *principal branch* and the values for this branch are called *principal values*.

Principal Values for Inverse Trigonometric Functions

Principal values for $x \geqq 0$	Principal values for $x < 0$
$0 \leqq \sin^{-1}x \leqq \pi/2$	$-\pi/2 \leqq \sin^{-1}x < 0$
$0 \leqq \cos^{-1}x \leqq \pi/2$	$\pi/2 < \cos^{-1}x \leqq \pi$
$0 \leqq \tan^{-1}x < \pi/2$	$-\pi/2 < \tan^{-1}x < 0$
$0 < \cot^{-1}x \leqq \pi/2$	$\pi/2 < \cot^{-1}x < \pi$
$0 \leqq \sec^{-1}x < \pi/2$	$\pi/2 < \sec^{-1}x \leqq \pi$
$0 < \csc^{-1}x \leqq \pi/2$	$-\pi/2 \leqq \csc^{-1}x < 0$

Relations Between Inverse Trigonometric Functions

In all cases, it is assumed that principal values are used.

10.59 $\sin^{-1}x + \cos^{-1}x = \pi/2$

10.60 $\tan^{-1}x + \cot^{-1}x = \pi/2$

10.61 $\sec^{-1}x + \csc^{-1}x = \pi/2$

10.62 $\csc^{-1}x = \sin^{-1}(1/x)$

10.63 $\sec^{-1}x = \cos^{-1}(1/x)$

10.64 $\cot^{-1}x = \tan^{-1}(1/x)$

10.65 $\sin^{-1}(-x) = -\sin^{-1}x$

10.66 $\cos^{-1}(-x) = \pi - \cos^{-1}x$

10.67 $\tan^{-1}(-x) = -\tan^{-1}x$

10.68 $\cot^{-1}(-x) = \pi - \cot^{-1}x$

10.69 $\sec^{-1}(-x) = \pi - \sec^{-1}x$

10.70 $\csc^{-1}(-x) = -\csc^{-1}x$

Graphs of Inverse Trigonometric Functions

In each graph, y is in radians. Solid portions of curves correspond to principal values.

10.71 $y = \sin^{-1}x$

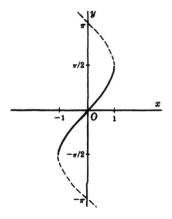

Figure 10-11

10.72 $y = \cos^{-1}x$

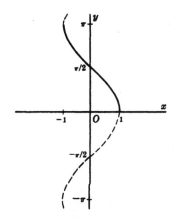

Figure 10-12

10.73 $y = \tan^{-1}x$

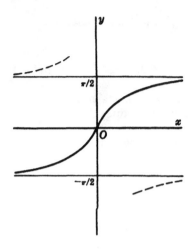

Figure 10-13

10.74 $y = \sin^{-1}x$

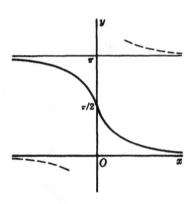

Figure 10-14

10.75 $y = \sec^{-1}x$

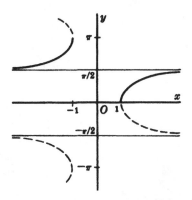

Figure 10-15

10.76 $y = \csc^{-1}x$

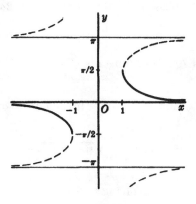

Figure 10-16

Relationships Between Sides and Angles of a Plane Triangle

The following results hold for any plane triangle ABC with sides a, b, c and angles A, B, C, as shown in Figure 10-17.

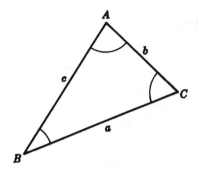

Figure 10-17

10.77 Law of Sines: $\dfrac{a}{\sin A} = \dfrac{b}{\sin B} = \dfrac{c}{\sin C}$

10.78 Law of Cosines: $c^2 = a^2 + b^2 - 2ab \cos C$

with similar relations involving the other sides and angles.

10.79 Law of Tangents: $\dfrac{a+b}{a-b} = \dfrac{\tan\frac{1}{2}(A+B)}{\tan\frac{1}{2}(A-B)}$

with similar relations involving the other sides and angles.

10.80 $\sin A = \dfrac{2}{bc}\sqrt{s(s-a)(s-b)(s-c)}$

where $s = \frac{1}{2}(a+b+c)$ is the semiperimeter of the triangle. Similar relations involving angles B and C can be obtained.

Relationships Between Sides and Angles of a Spherical Triangle

Sides a, b, c [which are arcs of great circles] are measured by their angles subtended at center O of the sphere. A, B, C are the angles opposite sides a, b, c respectively. Then the following results hold.

10.81 Law of Sines: $\dfrac{\sin a}{\sin A} = \dfrac{\sin b}{\sin B} = \dfrac{\sin c}{\sin C}$

10.82 Law of Cosines:
$$\cos a = \cos b \cos c + \sin b \sin c \cos A$$
$$\cos A = -\cos B \cos C + \sin B \sin C \cos a$$

with similar results involving other sides and angles.

10.83 Law of Tangents: $\dfrac{\tan \frac{1}{2}(A + B)}{\tan \frac{1}{2}(A - B)} = \dfrac{\tan \frac{1}{2}(a + b)}{\tan \frac{1}{2}(a - b)}$

with similar results involving other sides and angles.

10.84

$$\cos \frac{A}{2} = \sqrt{\frac{\sin s \sin (s - c)}{\sin b \sin c}}$$

where $s = \frac{1}{2}(a + b + c)$. **Similar results hold for other sides and angles.**

10.85

$$\cos \frac{a}{2} = \sqrt{\frac{\cos (S - B) \cos (S - C)}{\sin B \sin C}}$$

where $S = \frac{1}{2}(A + B + C)$. **Similar results hold for other sides and angles.**

11. Exponential and Logarithmic Functions

Laws of Exponents

In the following, p, q are real numbers, a, b are positive numbers and m, n are positive integers.

11.1 $a^p \cdot a^q = a^{p+q}$

11.2 $a^p/a^q = a^{p-q}$

11.3 $(a^p)^q = a^{pq}$

11.4 $a^0 = 1, a \neq 0$

11.5 $a^{-p} = 1/a^p$

11.6 $(ab)^p = a^p b^p$

11.7 $\sqrt[n]{a} = a^{1/n}$

11.8 $\sqrt[n]{a^m} = a^{m/n}$

11.9 $\sqrt[n]{a/b} = \sqrt[n]{a}/\sqrt[n]{b}$

In a^p, p is called the *exponent*, a is the *base* and a^p is the pth *power of a*. The function $y = a^x$ is called an *exponential function*.

Logarithms and Antilogarithms

In $a^p = N$ where $a \neq 0$ or 1, then $p = \log_a N$ is called the *logarithm* of N to the base a. The number $N = a^p$ is called the *antilogarithm* of p to the base a, written antilog$_a$ p.

Laws of Logarithms

11.10 $\log_a MN = \log_a M + \log_a N$

11.11 $\log_a \dfrac{M}{N} = \log_a M - \log_a N$

11.12 $\log_a M^p = p \log_a M$

Change of Base of Logarithms

The relationship between logarithms of a number N to different bases a and b is given by:

11.13 $\quad \log_a N = \dfrac{\log_b N}{\log_b a}$

In particular,

11.14 $\quad \log_e N = \ln N = 2.30258\ 50929\ 94\ldots \log_{10} N$

11.15 $\quad \log_{10} N = \log N = 0.43429\ 44819\ 03\ldots \log_e N$

Natural Logarithms and Antilogarithms

Natural logarithms and antilogarithms are those in which the base $a = e = 2.71828$. The natural logarithm of N is denoted by $\log_e N$ or $\ln N$.

Relationship Between Exponential and Trigonometric Functions

11.16
$$e^{i\theta} = \cos\theta + i\sin\theta, \quad e^{-i\theta} = \cos\theta - i\sin\theta$$

These are called *Euler's identities*. Here i is the imaginary unit.

11.17 $\quad \sin\theta = \dfrac{e^{i\theta} - e^{-i\theta}}{2i}$

11.18 $\quad \cos\theta = \dfrac{e^{i\theta} + e^{-i\theta}}{2}$

11.19 $\quad \tan\theta = \dfrac{e^{i\theta} - e^{-i\theta}}{i(e^{i\theta} + e^{-i\theta})} = -i\left(\dfrac{e^{i\theta} - e^{-i\theta}}{e^{i\theta} + e^{-i\theta}}\right)$

11.20 $\quad \cot\theta = i\left(\dfrac{e^{i\theta} + e^{-i\theta}}{e^{i\theta} - e^{-i\theta}}\right)$

11.21 $\quad \sec\theta = \dfrac{2}{e^{i\theta} + e^{-i\theta}}$

11.22 $\quad \csc\theta = \dfrac{2i}{e^{i\theta} - e^{-i\theta}}$

Periodicity of Exponential Functions

11.23 $\quad e^{i(\theta + 2k\pi)} = e^{i\theta} \qquad k = $ **integer**

From this, it is seen that e^x has period 2π.

Polar Form of Complex Numbers Expressed as an Exponential

The polar form of a complex number $z = x + iy$ can be written in terms of exponentials as follows:

11.24 $\quad z = x + iy = r(\cos\theta + i\sin\theta) = re^{i\theta}$

Operations with Complex Numbers in Polar Form

11.25 $\qquad (r_1 e^{i\theta_1})(r_2 e^{i\theta_2}) = r_1 r_2 e^{i(\theta_1 + \theta_2)}$

11.26 $\qquad \dfrac{r_1 e^{i\theta_1}}{r_2 e^{i\theta_2}} = \dfrac{r_1}{r_2} e^{i(\theta_1 - \theta_2)}$

11.27 $\quad (re^{i\theta})^p = r^p e^{ip\theta} \qquad$ [De Moivre's theorem]

11.28 $\quad (re^{i\theta})^{1/n} = [re^{i(\theta + 2k\pi)}]^{1/n} = r^{1/n} e^{i(\theta + 2k\pi)/n}$

Logarithms of a Complex Number

11.29 $\quad \ln(re^{i\theta}) = \ln r + i\theta + 2k\pi i \qquad k = $ **integer**

Section IV

CALCULUS

12. Derivatives

Definition of a Derivative

Suppose $y = f(x)$. The derivative of y or $f(x)$ is defined as

12.1 $\quad \dfrac{dy}{dx} = \lim_{h \to 0} \dfrac{f(x + h) - f(x)}{h} = \lim_{\Delta x \to 0} \dfrac{f(x + \Delta x) - f(x)}{\Delta x}$

where $h = \Delta x$. The derivative is also denoted by y', df/dx, or $f'(x)$. The process of taking a derivative is called *differentiation*.

General Rules of Differentiation

In the following, u, v, w are functions of x; a, b, c, n are constants; $e = 2.71828\ldots$ is the natural base of logarithms; $\ln u$ is the natural logarithm of u where it is assumed that $u > 0$; and all angles are in radians.

12.2 $\quad \dfrac{d}{dx}(c) = 0$

12.3 $\quad \dfrac{d}{dx}(cx) = c$

12.4 $\quad \dfrac{d}{dx}(cx^n) = ncx^{n-1}$

12.5 $\quad \dfrac{d}{dx}(u \pm v \pm w \pm \cdots) = \dfrac{du}{dx} \pm \dfrac{dv}{dx} \pm \dfrac{dw}{dx} \pm \cdots$

12.6 $\quad \dfrac{d}{dx}(cu) = c\,\dfrac{du}{dx}$

12.7 $\quad \dfrac{d}{dx}(uv) = u\,\dfrac{dv}{dx} + v\,\dfrac{du}{dx}$

12.8 $\quad \dfrac{d}{dx}(uvw) = uv\,\dfrac{dw}{dx} + uw\,\dfrac{dv}{dx} + vw\,\dfrac{du}{dx}$

12.9 $\quad \dfrac{d}{dx}\!\left(\dfrac{u}{v}\right) = \dfrac{v(du/dx) - u(dv/dx)}{v^2}$

12.10 $\quad \dfrac{d}{dx}(u^n) = nu^{n-1}\dfrac{du}{dx}$

12.11 $\quad \dfrac{dy}{dx} = \dfrac{dy}{du}\,\dfrac{du}{dx}$ $\quad$ (Chain rule)

12.12 $\quad \dfrac{du}{dx} = \dfrac{1}{dx/du}$

12.13 $\quad \dfrac{dy}{dx} = \dfrac{dy/du}{dx/du}$

Derivatives of Trigonometric and Inverse Trigonometric Functions

12.14 $\quad \dfrac{d}{dx}\sin u = \cos u\,\dfrac{du}{dx}$

12.15 $\quad \dfrac{d}{dx}\cos u = -\sin u\,\dfrac{du}{dx}$

12.16 $\quad \dfrac{d}{dx}\tan u = \sec^2 u\,\dfrac{du}{dx}$

12.17 $\quad \dfrac{d}{dx}\cot u = -\csc^2 u\,\dfrac{du}{dx}$

12.18 $\quad \dfrac{d}{dx}\sec u = \sec u \tan u\,\dfrac{du}{dx}$

12.19 $\quad \dfrac{d}{dx}\csc u = -\csc u \cot u\,\dfrac{du}{dx}$

12.20 $\quad \dfrac{d}{dx}\sin^{-1}u = \dfrac{1}{\sqrt{1-u^2}}\dfrac{du}{dx} \quad \left[-\dfrac{\pi}{2} < \sin^{-1}u < \dfrac{\pi}{2}\right]$

12.21 $\quad \dfrac{d}{dx}\cos^{-1}u = \dfrac{-1}{\sqrt{1-u^2}}\dfrac{du}{dx} \quad [0 < \cos^{-1}u < \pi]$

12.22 $\quad \dfrac{d}{dx}\tan^{-1}u = \dfrac{1}{1+u^2}\dfrac{du}{dx} \quad \left[-\dfrac{\pi}{2} < \tan^{-1}u < \dfrac{\pi}{2}\right]$

12.23 $\quad \dfrac{d}{dx}\cot^{-1}u = \dfrac{-1}{1+u^2}\dfrac{du}{dx} \quad [0 < \cot^{-1}u < \pi]$

12.24

$$\dfrac{d}{dx}\sec^{-1}u = \dfrac{1}{|u|\sqrt{u^2-1}}\dfrac{du}{dx} = \dfrac{\pm 1}{u\sqrt{u^2-1}}\dfrac{du}{dx} \quad \left[\begin{array}{l}+ \text{ if } 0 < \sec^{-1}u < \pi/2 \\ - \text{ if } \pi/2 < \sec^{-1}u < \pi\end{array}\right]$$

12.25

$$\dfrac{d}{dx}\csc^{-1}u = \dfrac{-1}{|u|\sqrt{u^2-1}}\dfrac{du}{dx} = \dfrac{\mp 1}{u\sqrt{u^2-1}}\dfrac{du}{dx} \quad \left[\begin{array}{l}- \text{ if } 0 < \csc^{-1}u < \pi/2 \\ + \text{ if } -\pi/2 < \csc^{-1}u < 0\end{array}\right]$$

Derivatives of Exponential and Logarithmic Functions

12.26 $\quad \dfrac{d}{dx}\log_a u = \dfrac{\log_a e}{u}\dfrac{du}{dx} \quad a \ne 0, 1$

12.27 $\quad \dfrac{d}{dx}\ln u = \dfrac{d}{dx}\log_e u = \dfrac{1}{u}\dfrac{du}{dx}$

12.28 $\quad \dfrac{d}{dx}a^u = a^u \ln a \dfrac{du}{dx}$

12.29 $\quad \dfrac{d}{dx}e^u = e^u \dfrac{du}{dx}$

12.30 $\quad \dfrac{d}{dx}u^v = \dfrac{d}{dx}e^{v\ln u} = e^{v\ln u}\dfrac{d}{dx}[v\ln u] = vu^{v-1}\dfrac{du}{dx} + u^v \ln u \dfrac{dv}{dx}$

Higher Derivatives

The second, third, and higher derivatives are defined as follows:

12.31   **Second derivative** $= \dfrac{d}{dx}\left(\dfrac{dy}{dx}\right) = \dfrac{d^2y}{dx^2} = f''(x) = y''$

12.32   **Third derivative** $= \dfrac{d}{dx}\left(\dfrac{d^2y}{dx^2}\right) = \dfrac{d^3y}{dx^3} = f'''(x) = y'''$

12.33   **nth derivative** $= \dfrac{d}{dx}\left(\dfrac{d^{n-1}y}{dx^{n-1}}\right) = \dfrac{d^ny}{dx^n} = f^{(n)}(x) = y^{(n)}$

Differentials

Let $y = f(x)$ and $\Delta y = f(x + \Delta x) - f(x)$. Then

12.34 $\dfrac{\Delta y}{\Delta x} = \dfrac{f(x + \Delta x) - f(x)}{\Delta x} = f'(x) + \epsilon = \dfrac{dy}{dx} + \epsilon$

where $\epsilon \to 0$ as $\Delta x \to 0$. Thus,

12.35 $\Delta y = f'(x)\Delta x + \epsilon\,\Delta x$

If we call $\Delta x = dx$ the differential of x, then we define the differential of y to be

12.36 $dy = f'(x)\,dx$

The rules for differentials are exactly analogous to those for derivatives. As examples, we observe that:

12.37 $d(u \pm v \pm w \pm \cdots) = du \pm dv \pm dw \pm \cdots$

12.38 $d(uv) = u\,dv + v\,du$

12.39 $d\left(\dfrac{u}{v}\right) = \dfrac{v\,du - u\,dv}{v^2}$

12.40 $d(u^n) = nu^{n-1}\,du$

12.41 $d(\sin u) = \cos u\,du$

12.42 $d(\cos u) = -\sin u\, du$

13. Indefinite Integrals

Definition of an Indefinite Integral

If $\dfrac{dy}{dx} = f(x)$, then y is the function whose derivative is $f(x)$ and is called the antiderivative of $f(x)$ or the *indefinate integral of f(x)*, denoted by $\displaystyle\int f(x)\, dx$. Similarly, if $y = \displaystyle\int f(u)\, du$, then $\dfrac{dy}{du} = f(u)$.

Since the derivative of a constant is zero, all indefinate integrals differ by an arbitrary constant.

General Rules of Integration

In the following, u, v, w are functions of x; a, b, c, n are constants; $e = 2.71828\ldots$ is the natural base of logarithms; ln u is the natural logarithm of u where it is assumed that $u > 0$; all angles are in radians; and all constants of integration are omitted but implied.

13.1 $\displaystyle\int a\, dx = ax$

13.2 $\displaystyle\int af(x)\, dx = a \int f(x)\, dx$

13.3 $\displaystyle\int (u \pm v \pm w \pm \cdots)\, dx = \int u\, dx \pm \int v\, dx \pm \int w\, dx \pm \cdots$

13.4 $\displaystyle\int u\, dv = uv - \int v\, du$ [Integration by parts]

13.5 $\displaystyle\int f(ax)\, dx = \frac{1}{a} \int f(u)\, du$

13.6 $\displaystyle\int F\{f(x)\}\, dx = \int F(u)\frac{dx}{du}\, du = \int \frac{F(u)}{f'(x)}\, du$ where $u = f(x)$

13.7 $\quad \displaystyle\int u^n \, du = \frac{u^{n+1}}{n+1}, \quad n \neq -1 \quad$ [For $n = $ -1, see 13.8]

13.8 $\quad \displaystyle\int \frac{du}{u} = \ln u \quad$ if $u > 0$ or $\ln(-u)$ if $u < 0$

$\qquad\qquad = \ln|u|$

13.9 $\quad \displaystyle\int e^u \, du = e^u$

13.10 $\quad \displaystyle\int a^u \, du = \int e^{u \ln a} \, du = \frac{e^{u \ln a}}{\ln a} = \frac{a^u}{\ln a}, \quad a > 0, \, a \neq 1$

13.11 $\quad \displaystyle\int \sin u \, du = -\cos u$

13.12 $\quad \displaystyle\int \cos u \, du = \sin u$

13.13 $\quad \displaystyle\int \tan u \, du = \ln \sec u = -\ln \cos u$

13.14 $\quad \displaystyle\int \cot u \, du = \ln \sin u$

13.15 $\quad \displaystyle\int \sec u \, du = \ln(\sec u + \tan u) = \ln \tan \left(\frac{u}{2} + \frac{\pi}{4} \right)$

13.16 $\quad \displaystyle\int \csc u \, du = \ln(\csc u - \cot u) = \ln \tan \frac{u}{2}$

13.17 $\quad \displaystyle\int \sec^2 u \, du = \tan u$

13.18 $\quad \displaystyle\int \csc^2 u \, du = -\cot u$

13.19 $\quad \displaystyle\int \tan^2 u \, du = \tan u - u$

13.20 $\quad \displaystyle\int \cot^2 u \, du = -\cot u - u$

13.21 $\quad \displaystyle\int \sin^2 u \, du = \frac{u}{2} - \frac{\sin 2u}{4} = \tfrac{1}{2}(u - \sin u \cos u)$

13.22 $\displaystyle\int \cos^2 u \, du = \frac{u}{2} + \frac{\sin 2u}{4} = \tfrac{1}{2}(u + \sin u \cos u)$

13.23 $\displaystyle\int \frac{du}{u^2 + a^2} = \frac{1}{a}\tan^{-1}\frac{u}{a}$

13.24 $\displaystyle\int \frac{du}{u^2 - a^2} = \frac{1}{2a}\ln\left(\frac{u-a}{u+a}\right) = -\frac{1}{a}\coth^{-1}\frac{u}{a} \quad u^2 > a^2$

13.25 $\displaystyle\int \frac{du}{a^2 - u^2} = \frac{1}{2a}\ln\left(\frac{a+u}{a-u}\right) = \frac{1}{a}\tanh^{-1}\frac{u}{a} \quad u^2 < a^2$

13.26 $\displaystyle\int \frac{du}{\sqrt{a^2 - u^2}} = \sin^{-1}\frac{u}{a}$

13.27 $\displaystyle\int \frac{du}{\sqrt{u^2 + a^2}} = \ln(u + \sqrt{u^2 + a^2}) \quad$ or $\quad \sinh^{-1}\frac{u}{a}$

13.28 $\displaystyle\int \frac{du}{\sqrt{u^2 - a^2}} = \ln(u + \sqrt{u^2 - a^2})$

13.29 $\displaystyle\int \frac{du}{u\sqrt{u^2 - a^2}} = \frac{1}{a}\sec^{-1}\left|\frac{u}{a}\right|$

13.30 $\displaystyle\int \frac{du}{u\sqrt{u^2 + a^2}} = -\frac{1}{a}\ln\left(\frac{a + \sqrt{u^2 + a^2}}{u}\right)$

13.31 $\displaystyle\int \frac{du}{u\sqrt{a^2 - u^2}} = -\frac{1}{a}\ln\left(\frac{a + \sqrt{a^2 - u^2}}{u}\right)$

Important Transformations

Often in practice an integral can be simplified by using an appropriate transformation or substitution. The following list gives some transformations and their effects.

13.32 $\displaystyle\int F(ax + b)\, dx = \frac{1}{a}\int F(u)\, du$

where $\quad u = ax + b$

13.33 $\displaystyle\int F(\sqrt{ax+b})\,dx = \frac{2}{a}\int u\,F(u)\,du$

  where $u = \sqrt{ax+b}$

13.34 $\displaystyle\int F(\sqrt{ax+b})\,dx = \frac{2}{a}\int u\,F(u)\,du$

  where $u = \sqrt{ax+b}$

13.35 $\displaystyle\int F(\sqrt{ax+b})\,dx = \frac{2}{a}\int u\,F(u)\,du$

  where $u = \sqrt{ax+b}$

13.36 $\displaystyle\int F(\sqrt{ax+b})\,dx = \frac{2}{a}\int u\,F(u)\,du$

  where $u = \sqrt{ax+b}$

13.37 $\displaystyle\int F(\sqrt{ax+b})\,dx = \frac{2}{a}\int u\,F(u)\,du$

  where $u = \sqrt{ax+b}$

13.38 $\displaystyle\int F(e^{ax})\,dx = \frac{1}{a}\int \frac{F(u)}{u}\,du$

  where $u = e^{ax}$

13.39 $\displaystyle\int F(\sqrt{ax+b})\,dx = \frac{2}{a}\int u\,F(u)\,du$

  where $u = \sqrt{ax+b}$

13.40 $\displaystyle\int F\left(\sin^{-1}\frac{x}{a}\right)dx = a\int F(u)\cos u\,du$

  where $u = \sin^{-1}\dfrac{x}{a}$

13.41 $\displaystyle\int F(\sin x,\ \cos x)\,dx = 2\int F\left(\frac{2u}{1+u^2},\frac{1-u^2}{1+u^2}\right)\frac{du}{1+u^2}$

  where $u = \tan\dfrac{x}{2}$

14. Tables of Special Indefinite Integrals

Integrals Involving $ax + b$

14.1 $\displaystyle\int \frac{dx}{ax+b} = \frac{1}{a}\ln(ax+b)$

14.2 $\displaystyle\int \frac{x\,dx}{ax+b} = \frac{x}{a} - \frac{b}{a^2}\ln(ax+b)$

14.3 $\displaystyle\int \frac{x^2\,dx}{ax+b} = \frac{(ax+b)^2}{2a^3} - \frac{2b(ax+b)}{a^3} + \frac{b^2}{a^3}\ln(ax+b)$

14.4 $\displaystyle\int \frac{dx}{x(ax+b)} = \frac{1}{b}\ln\left(\frac{x}{ax+b}\right)$

14.5 $\displaystyle\int \frac{dx}{x^2(ax+b)} = -\frac{1}{bx} + \frac{a}{b^2}\ln\left(\frac{ax+b}{x}\right)$

14.6 $\displaystyle\int \frac{dx}{(ax+b)^2} = \frac{-1}{a(ax+b)}$

14.7 $\displaystyle\int \frac{x\,dx}{(ax+b)^2} = \frac{b}{a^2(ax+b)} + \frac{1}{a^2}\ln(ax+b)$

14.8 $\displaystyle\int \frac{x^2\,dx}{(ax+b)^2} = \frac{ax+b}{a^3} - \frac{b^2}{a^3(ax+b)} - \frac{2b}{a^3}\ln(ax+b)$

14.9 $\displaystyle\int (ax+b)^n\,dx = \frac{(ax+b)^{n+1}}{(n+1)a}.$

If $n = -1$, see 14.1.

14.10 $\displaystyle\int x(ax+b)^n\,dx = \frac{(ax+b)^{n+2}}{(n+2)a^2} - \frac{b(ax+b)^{n+1}}{(n+1)a^2}, \quad n \neq -1, -2$

If $n = -1, -2$, see 14.2 and 14.7.

14.11 $\displaystyle\int x^2(ax+b)^n\,dx = \frac{(ax+b)^{n+3}}{(n+3)a^3} - \frac{2b(ax+b)^{n+2}}{(n+2)a^3} + \frac{b^2(ax+b)^{n+1}}{(n+1)a^3}$

If $n = -1, -2$, see 14.3 and 14.8.

14.12

$$\int x^m(ax+b)^n\,dx = \begin{cases} \dfrac{x^{m+1}(ax+b)^n}{m+n+1} + \dfrac{nb}{m+n+1}\displaystyle\int x^m(ax+b)^{n-1}\,dx \\[3mm] \dfrac{x^m(ax+b)^{n+1}}{(m+n+1)a} - \dfrac{mb}{(m+n+1)a}\displaystyle\int x^{m-1}(ax+b)^n\,dx \\[3mm] \dfrac{-x^{m+1}(ax+b)^{n+1}}{(n+1)b} + \dfrac{m+n+2}{(n+1)b}\displaystyle\int x^m(ax+b)^{n+1}\,dx \end{cases}$$

Integrals Involving $\sqrt{ax+b}$

14.13 $\displaystyle\int \frac{dx}{\sqrt{ax+b}} = \frac{2\sqrt{ax+b}}{a}$

14.14 $\displaystyle\int \frac{x\,dx}{\sqrt{ax+b}} = \frac{2(ax-2b)}{3a^2}\sqrt{ax+b}$

14.15 $\displaystyle\int \frac{x^2\,dx}{\sqrt{ax+b}} = \frac{2(3a^2x^2-4abx+8b^2)}{15a^3}\sqrt{ax+b}$

14.16 $\displaystyle\int \frac{dx}{x\sqrt{ax+b}} = \begin{cases} \dfrac{1}{\sqrt{b}}\ln\left(\dfrac{\sqrt{ax+b}-\sqrt{b}}{\sqrt{ax+b}+\sqrt{b}}\right) \\[3mm] \dfrac{2}{\sqrt{-b}}\tan^{-1}\sqrt{\dfrac{ax+b}{-b}} \end{cases}$

14.17 $\displaystyle\int \frac{dx}{x^2\sqrt{ax+b}} = -\frac{\sqrt{ax+b}}{bx} - \frac{a}{2b}\int \frac{dx}{x\sqrt{ax+b}}$

[See 14.24].

14.18 $\displaystyle\int \sqrt{ax+b}\, dx = \frac{2\sqrt{(ax+b)^3}}{3a}$

14.19 $\displaystyle\int x\sqrt{ax+b}\, dx = \frac{2(3ax-2b)}{15a^2}\sqrt{(ax+b)^3}$

14.20 $\displaystyle\int x^2\sqrt{ax+b}\, dx = \frac{2(15a^2x^2-12abx+8b^2)}{105a^3}\sqrt{(ax+b)^3}$

14.21 $\displaystyle\int \frac{\sqrt{ax+b}}{x}\, dx = 2\sqrt{ax+b}+b\int \frac{dx}{x\sqrt{ax+b}}$

[See 14.24].

14.22 $\displaystyle\int \frac{\sqrt{ax+b}}{x^2}\, dx = -\frac{\sqrt{ax+b}}{x}+\frac{a}{2}\int \frac{dx}{x\sqrt{ax+b}}$

[See 14.24].

14.23 $\displaystyle\int \frac{x^m}{\sqrt{ax+b}}\, dx = \frac{2x^m\sqrt{ax+b}}{(2m+1)a}-\frac{2mb}{(2m+1)a}\int \frac{x^{m-1}}{\sqrt{ax+b}}\, dx$

14.24 $\displaystyle\int \frac{dx}{x^m\sqrt{ax+b}} = -\frac{\sqrt{ax+b}}{(m-1)bx^{m-1}}-\frac{(2m-3)a}{(2m-2)b}\int \frac{dx}{x^{m-1}\sqrt{ax+b}}$

14.25

$\displaystyle\int x^m\sqrt{ax+b}\, dx = \frac{2x^m}{(2m+3)a}(ax+b)^{3/2}-\frac{2mb}{(2m+3)a}\int x^{m-1}\sqrt{ax+b}\, dx$

14.26 $\displaystyle\int \frac{\sqrt{ax+b}}{x^m}\, dx = -\frac{\sqrt{ax+b}}{(m-1)x^{m-1}}+\frac{a}{2(m-1)}\int \frac{dx}{x^{m-1}\sqrt{ax+b}}$

14.27 $\displaystyle\int \frac{\sqrt{ax+b}}{x^m}\, dx = \frac{-(ax+b)^{3/2}}{(m-1)bx^{m-1}}-\frac{(2m-5)a}{(2m-2)b}\int \frac{\sqrt{ax+b}}{x^{m-1}}\, dx$

14.28 $\displaystyle\int (ax+b)^{m/2}\, dx = \frac{2(ax+b)^{(m+2)/2}}{a^2(m+2)}$

14.29 $\int x(ax+b)^{m/2}\,dx = \dfrac{2(ax+b)^{(m+4)/2}}{a^2(m+4)} - \dfrac{2b(ax+b)^{(m+2)/2}}{a^2(m+2)}$

14.30

$$\int x^2(ax+b)^{m/2}\,dx = \frac{2(ax+b)^{(m+6)/2}}{a^3(m+6)} - \frac{4b(ax+b)^{(m+4)/2}}{a^3(m+4)} + \frac{2b^2(ax+b)^{(m+2)/2}}{a^3(m+2)}$$

14.31 $\displaystyle\int \frac{(ax+b)^{m/2}}{x}\,dx = \frac{2(ax+b)^{m/2}}{m} + b\int \frac{(ax+b)^{(m-2)/2}}{x}\,dx$

14.32 $\displaystyle\int \frac{(ax+b)^{m/2}}{x^2}\,dx = -\frac{(ax+b)^{(m+2)/2}}{bx} + \frac{ma}{2b}\int \frac{(ax+b)^{m/2}}{x}\,dx$

14.33 $\displaystyle\int \frac{dx}{x(ax+b)^{m/2}} = \frac{2}{(m-2)b(ax+b)^{(m-2)/2}} + \frac{1}{b}\int \frac{dx}{x(ax+b)^{(m-2)/2}}$

Integrals Involving $\sqrt{ax+b}$ and $\sqrt{px+q}$

14.34 $\displaystyle\int \frac{dx}{(ax+b)(px+q)} = \frac{1}{bp-aq}\ln\left(\frac{px+q}{ax+b}\right)$

14.35 $\displaystyle\int \frac{x\,dx}{(ax+b)(px+q)} = \frac{1}{bp-aq}\left\{\frac{b}{a}\ln(ax+b) - \frac{q}{p}\ln(px+q)\right\}$

14.36 $\displaystyle\int \frac{dx}{(ax+b)^2(px+q)} = \frac{1}{bp-aq}\left\{\frac{1}{ax+b} + \frac{p}{bp-aq}\ln\left(\frac{px+q}{ax+b}\right)\right\}$

14.37 $\displaystyle\int \frac{x\,dx}{(ax+b)^2(px+q)} = \frac{1}{bp-aq}\left\{\frac{q}{bp-aq}\ln\left(\frac{ax+b}{px+q}\right) - \frac{b}{a(ax+b)}\right\}$

14.38 $\displaystyle\int \frac{x^2\,dx}{(ax+b)^2(px+q)} = \frac{b^2}{(bp-aq)a^2(ax+b)} +$

$\qquad\qquad \dfrac{1}{(bp-aq)^2}\left\{\dfrac{q^2}{p}\ln(px+q) + \dfrac{b(bp-2aq)}{a^2}\ln(ax+b)\right\}$

14.39

$$\int \frac{dx}{(ax+b)^m(px+q)^n} = \frac{-1}{(n-1)(bp-aq)}$$

$$\left\{ \frac{1}{(ax+b)^{m-1}(px+q)^{n-1}} + a(m+n-2) \int \frac{dx}{(ax+b)^m(px+q)^{n-1}} \right\}$$

14.40 $\displaystyle\int \frac{ax+b}{px+q}\,dx = \frac{ax}{p} + \frac{bp-aq}{p^2}\ln(px+q)$

14.41

$$\int \frac{(ax+b)^m}{(px+q)^n}\,dx =$$

$$\begin{cases} \dfrac{-1}{(n-1)(bp-aq)}\left\{ \dfrac{(ax+b)^{m+1}}{(px+q)^{n-1}} + (n-m-2)a \int \dfrac{(ax+b)^m}{(px+q)^{n-1}}\,dx \right\} \\[4mm] \dfrac{-1}{(n-m-1)p}\left\{ \dfrac{(ax+b)^m}{(px+q)^{n-1}} + m(bp-aq) \int \dfrac{(ax+b)^{m-1}}{(px+q)^n}\,dx \right\} \\[4mm] \dfrac{-1}{(n-1)p}\left\{ \dfrac{(ax+b)^m}{(px+q)^{n-1}} - ma \int \dfrac{(ax+b)^{m-1}}{(px+q)^{n-1}}\,dx \right\} \end{cases}$$

Integrals Involving $\sqrt{ax+b}$ and $\sqrt{px+q}$

14.42 $\displaystyle\int \frac{dx}{\sqrt{(ax+b)(px+q)}} = \begin{cases} \dfrac{2}{\sqrt{ap}}\ln\left(\sqrt{a(px+q)} + \sqrt{p(ax+b)}\right) \\[4mm] \dfrac{2}{\sqrt{-ap}}\tan^{-1}\sqrt{\dfrac{-p(ax+b)}{a(px+q)}} \end{cases}$

14.43

$$\int \frac{x\,dx}{\sqrt{(ax+b)(px+q)}} = \frac{\sqrt{(ax+b)(px+q)}}{ap} - \frac{bp+aq}{2ap} \int \frac{dx}{\sqrt{(ax+b)(px+q)}}$$

14.44 $\displaystyle\int \sqrt{(ax+b)(px+q)}\,dx = \frac{2apx+bp+aq}{4ap}\sqrt{(ax+b)(px+q)} -$

$$\frac{(bp-aq)^2}{8ap} \int \frac{dx}{\sqrt{(ax+b)(px+q)}}$$

14.45 $$\int \sqrt{\frac{px+q}{ax+b}}\,dx = \frac{\sqrt{(ax+b)(px+q)}}{a} + \frac{aq-bp}{2a}\int \frac{dx}{\sqrt{(ax+b)(px+q)}}$$

14.46 $$\int \frac{dx}{(px+q)\sqrt{(ax+b)(px+q)}} = \frac{2\sqrt{ax+b}}{(aq-bp)\sqrt{px+q}}$$

Integrals Involving $x^2 + a^2$

14.47 $$\int \frac{dx}{x^2+a^2} = \frac{1}{a}\tan^{-1}\frac{x}{a}$$

14.48 $$\int \frac{x\,dx}{x^2+a^2} = \frac{1}{2}\ln(x^2+a^2)$$

14.49 $$\int \frac{x^2\,dx}{x^2+a^2} = x - a\tan^{-1}\frac{x}{a}$$

14.50 $$\int \frac{dx}{x(x^2+a^2)} = \frac{1}{2a^2}\ln\left(\frac{x^2}{x^2+a^2}\right)$$

14.51 $$\int \frac{dx}{x^2(x^2+a^2)} = -\frac{1}{a^2x} - \frac{1}{a^3}\tan^{-1}\frac{x}{a}$$

14.52 $$\int \frac{dx}{(x^2+a^2)^2} = \frac{x}{2a^2(x^2+a^2)} + \frac{1}{2a^3}\tan^{-1}\frac{x}{a}$$

14.53 $$\int \frac{x\,dx}{(x^2+a^2)^2} = \frac{-1}{2(x^2+a^2)}$$

14.54 $$\int \frac{x^2\,dx}{(x^2+a^2)^2} = \frac{-x}{2(x^2+a^2)} + \frac{1}{2a}\tan^{-1}\frac{x}{a}$$

14.55 $$\int \frac{dx}{x(x^2+a^2)^2} = \frac{1}{2a^2(x^2+a^2)} + \frac{1}{2a^4}\ln\left(\frac{x^2}{x^2+a^2}\right)$$

14.56 $$\int \frac{dx}{x^2(x^2+a^2)^2} = -\frac{1}{a^4x} - \frac{x}{2a^4(x^2+a^2)} - \frac{3}{2a^5}\tan^{-1}\frac{x}{a}$$

14.57 $$\int \frac{dx}{(x^2+a^2)^n} = \frac{x}{2(n-1)a^2(x^2+a^2)^{n-1}} + \frac{2n-3}{(2n-2)a^2} \int \frac{dx}{(x^2+a^2)^{n-1}}$$

14.58 $$\int \frac{x\,dx}{(x^2+a^2)^n} = \frac{-1}{2(n-1)(x^2+a^2)^{n-1}}$$

14.59 $$\int \frac{dx}{x(x^2+a^2)^n} = \frac{1}{2(n-1)a^2(x^2+a^2)^{n-1}} + \frac{1}{a^2} \int \frac{dx}{x(x^2+a^2)^{n-1}}$$

14.60 $$\int \frac{x^m\,dx}{(x^2+a^2)^n} = \int \frac{x^{m-2}\,dx}{(x^2+a^2)^{n-1}} - a^2 \int \frac{x^{m-2}\,dx}{(x^2+a^2)^n}$$

14.61 $$\int \frac{dx}{x^m(x^2+a^2)^n} = \frac{1}{a^2} \int \frac{dx}{x^m(x^2+a^2)^{n-1}} - \frac{1}{a^2} \int \frac{dx}{x^{m-2}(x^2+a^2)^n}$$

Integrals Involving $x^2 - a^2$, $x^2 > a^2$

14.62 $$\int \frac{dx}{x^2-a^2} = \frac{1}{2a} \ln\left(\frac{x-a}{x+a}\right)$$

14.63 $$\int \frac{x\,dx}{x^2-a^2} = \frac{1}{2} \ln(x^2-a^2)$$

14.64 $$\int \frac{x^2\,dx}{x^2-a^2} = x + \frac{a}{2} \ln\left(\frac{x-a}{x+a}\right)$$

14.65 $$\int \frac{dx}{x(x^2-a^2)} = \frac{1}{2a^2} \ln\left(\frac{x^2-a^2}{x^2}\right)$$

14.66 $$\int \frac{dx}{x^2(x^2-a^2)} = \frac{1}{a^2x} + \frac{1}{2a^3} \ln\left(\frac{x-a}{x+a}\right)$$

14.67 $$\int \frac{dx}{(x^2-a^2)^2} = \frac{-x}{2a^2(x^2-a^2)} - \frac{1}{4a^3} \ln\left(\frac{x-a}{x+a}\right)$$

14.68 $$\int \frac{x\,dx}{(x^2-a^2)^2} = \frac{-1}{2(x^2-a^2)}$$

14.69 $$\int \frac{x^2\,dx}{(x^2-a^2)^2} = \frac{-x}{2(x^2-a^2)} + \frac{1}{4a} \ln\left(\frac{x-a}{x+a}\right)$$

14.70 $\displaystyle \int \frac{dx}{x(x^2-a^2)^n} = \frac{-1}{2(n-1)a^2(x^2-a^2)^{n-1}} - \frac{1}{a^2}\int \frac{dx}{x(x^2-a^2)^{n-1}}$

14.71 $\displaystyle \int \frac{x^m\,dx}{(x^2-a^2)^n} = \int \frac{x^{m-2}\,dx}{(x^2-a^2)^{n-1}} + a^2 \int \frac{x^{m-2}\,dx}{(x^2-a^2)^n}$

14.72 $\displaystyle \int \frac{dx}{x^m(x^2-a^2)^n} = \frac{1}{a^2}\int \frac{dx}{x^{m-2}(x^2-a^2)^n} - \frac{1}{a^2}\int \frac{dx}{x^m(x^2-a^2)^{n-1}}$

14.73 $\displaystyle \int \frac{dx}{x(x^2-a^2)^2} = \frac{-1}{2a^2(x^2-a^2)} + \frac{1}{2a^4}\ln\left(\frac{x^2}{x^2-a^2}\right)$

14.74 $\displaystyle \int \frac{dx}{x^2(x^2-a^2)^2} = -\frac{1}{a^4x} - \frac{x}{2a^4(x^2-a^2)} - \frac{3}{4a^5}\ln\left(\frac{x-a}{x+a}\right)$

14.75 $\displaystyle \int \frac{dx}{(x^2-a^2)^n} = \frac{-x}{2(n-1)a^2(x^2-a^2)^{n-1}} - \frac{2n-3}{(2n-2)a^2}\int \frac{dx}{(x^2-a^2)^{n-1}}$

14.76 $\displaystyle \int \frac{x\,dx}{(x^2-a^2)^n} = \frac{-1}{2(n-1)(x^2-a^2)^{n-1}}$

Integrals Involving $a^2 - x^2$, $x^2 < a^2$

14.77 $\displaystyle \int \frac{dx}{a^2-x^2} = \frac{1}{2a}\ln\left(\frac{a+x}{a-x}\right)$

14.78 $\displaystyle \int \frac{x\,dx}{a^2-x^2} = -\frac{1}{2}\ln(a^2-x^2)$

14.79 $\displaystyle \int \frac{x^2\,dx}{a^2-x^2} = -x + \frac{a}{2}\ln\left(\frac{a+x}{a-x}\right)$

14.80 $\displaystyle \int \frac{dx}{x(a^2-x^2)} = \frac{1}{2a^2}\ln\left(\frac{x^2}{a^2-x^2}\right)$

14.81 $\displaystyle \int \frac{dx}{x^2(a^2-x^2)} = -\frac{1}{a^2x} + \frac{1}{2a^3}\ln\left(\frac{a+x}{a-x}\right)$

14.82 $\displaystyle\int \frac{dx}{(a^2 - x^2)^2} = \frac{x}{2a^2(a^2 - x^2)} + \frac{1}{4a^3}\ln\left(\frac{a+x}{a-x}\right)$

14.83 $\displaystyle\int \frac{x\,dx}{(a^2 - x^2)^2} = \frac{1}{2(a^2 - x^2)}$

14.84 $\displaystyle\int \frac{x^2\,dx}{(a^2 - x^2)^2} = \frac{x}{2(a^2 - x^2)} - \frac{1}{4a}\ln\left(\frac{a+x}{a-x}\right)$

14.85 $\displaystyle\int \frac{dx}{x(a^2 - x^2)^2} = \frac{1}{2a^2(a^2 - x^2)} + \frac{1}{2a^4}\ln\left(\frac{x^2}{a^2 - x^2}\right)$

14.86 $\displaystyle\int \frac{dx}{x^2(a^2 - x^2)^2} = \frac{-1}{a^4 x} + \frac{x}{2a^4(a^2 - x^2)} + \frac{3}{4a^5}\ln\left(\frac{a+x}{a-x}\right)$

14.87 $\displaystyle\int \frac{dx}{(a^2 - x^2)^n} = \frac{x}{2(n-1)a^2(a^2 - x^2)^{n-1}} + \frac{2n-3}{(2n-2)a^2}\int \frac{dx}{(a^2 - x^2)^{n-1}}$

14.88 $\displaystyle\int \frac{x\,dx}{(a^2 - x^2)^n} = \frac{1}{2(n-1)(a^2 - x^2)^{n-1}}$

14.89 $\displaystyle\int \frac{dx}{x(a^2 - x^2)^n} = \frac{1}{2(n-1)a^2(a^2 - x^2)^{n-1}} + \frac{1}{a^2}\int \frac{dx}{x(a^2 - x^2)^{n-1}}$

14.90 $\displaystyle\int \frac{x^m\,dx}{(a^2 - x^2)^n} = a^2\int \frac{x^{m-2}\,dx}{(a^2 - x^2)^n} - \int \frac{x^{m-2}\,dx}{(a^2 - x^2)^{n-1}}$

14.91 $\displaystyle\int \frac{dx}{x^m(a^2 - x^2)^n} = \frac{1}{a^2}\int \frac{dx}{x^m(a^2 - x^2)^{n-1}} + \frac{1}{a^2}\int \frac{dx}{x^{m-2}(a^2 - x^2)^n}$

Integrals Involving $\sqrt{x^2 + a^2}$

14.92 $\displaystyle\int \frac{dx}{\sqrt{x^2 + a^2}} = \ln(x + \sqrt{x^2 + a^2})$

14.93 $\displaystyle\int \frac{x\,dx}{\sqrt{x^2 + a^2}} = \sqrt{x^2 + a^2}$

14.94 $\displaystyle\int \frac{x^2\,dx}{\sqrt{x^2 + a^2}} = \frac{x\sqrt{x^2 + a^2}}{2} - \frac{a^2}{2}\ln(x + \sqrt{x^2 + a^2})$

14.95 $\quad \displaystyle\int \frac{dx}{x\sqrt{x^2+a^2}} = -\frac{1}{a}\ln\left(\frac{a+\sqrt{x^2+a^2}}{x}\right)$

14.96 $\quad \displaystyle\int \frac{dx}{x^2\sqrt{x^2+a^2}} = -\frac{\sqrt{x^2+a^2}}{a^2 x}$

14.97 $\quad \displaystyle\int \sqrt{x^2+a^2}\,dx = \frac{x\sqrt{x^2+a^2}}{2} + \frac{a^2}{2}\ln(x+\sqrt{x^2+a^2})$

14.98 $\quad \displaystyle\int x\sqrt{x^2+a^2}\,dx = \frac{(x^2+a^2)^{3/2}}{3}$

14.99 $\quad \displaystyle\int x^2\sqrt{x^2+a^2}\,dx = \frac{x(x^2+a^2)^{3/2}}{4} - \frac{a^2 x\sqrt{x^2+a^2}}{8} - \frac{a^4}{8}\ln(x+\sqrt{x^2+a^2})$

14.100 $\quad \displaystyle\int \frac{\sqrt{x^2+a^2}}{x}\,dx = \sqrt{x^2+a^2} - a\ln\left(\frac{a+\sqrt{x^2+a^2}}{x}\right)$

14.101 $\quad \displaystyle\int \frac{\sqrt{x^2+a^2}}{x^2}\,dx = -\frac{\sqrt{x^2+a^2}}{x} + \ln(x+\sqrt{x^2+a^2})$

Integrals Involving $\sqrt{x^2 - a^2}$

14.102 $\quad \displaystyle\int \frac{dx}{\sqrt{x^2-a^2}} = \ln(x+\sqrt{x^2-a^2}), \qquad \displaystyle\int \frac{x\,dx}{\sqrt{x^2-a^2}} = \sqrt{x^2-a^2}$

14.103 $\quad \displaystyle\int \frac{x^2\,dx}{\sqrt{x^2-a^2}} = \frac{x\sqrt{x^2-a^2}}{2} + \frac{a^2}{2}\ln(x+\sqrt{x^2-a^2})$

14.104 $\quad \displaystyle\int \frac{dx}{x\sqrt{x^2-a^2}} = \frac{1}{a}\sec^{-1}\left|\frac{x}{a}\right|$

14.105 $\quad \displaystyle\int \frac{dx}{x^2\sqrt{x^2-a^2}} = \frac{\sqrt{x^2-a^2}}{a^2 x}$

14.106 $\quad \displaystyle\int \sqrt{x^2-a^2}\,dx = \frac{x\sqrt{x^2-a^2}}{2} - \frac{a^2}{2}\ln(x+\sqrt{x^2-a^2})$

14.107 $\quad \displaystyle\int x\sqrt{x^2-a^2}\,dx = \frac{(x^2-a^2)^{3/2}}{3}$

14.108 $\displaystyle\int x^2\sqrt{x^2-a^2}\,dx = \frac{x(x^2-a^2)^{3/2}}{4} + \frac{a^2x\sqrt{x^2-a^2}}{8} - \frac{a^4}{8}\ln\left(x+\sqrt{x^2-a^2}\right)$

14.109 $\displaystyle\int \frac{\sqrt{x^2-a^2}}{x}\,dx = \sqrt{x^2-a^2} - a\sec^{-1}\left|\frac{x}{a}\right|$

14.110 $\displaystyle\int \frac{\sqrt{x^2-a^2}}{x^2}\,dx = -\frac{\sqrt{x^2-a^2}}{x} + \ln\left(x+\sqrt{x^2-a^2}\right)$

Integrals Involving $\sqrt{a^2-x^2}$

14.111 $\displaystyle\int \frac{dx}{\sqrt{a^2-x^2}} = \sin^{-1}\frac{x}{a}$

14.112 $\displaystyle\int \frac{x\,dx}{\sqrt{a^2-x^2}} = -\sqrt{a^2-x^2}$

14.113 $\displaystyle\int \frac{x^2\,dx}{\sqrt{a^2-x^2}} = -\frac{x\sqrt{a^2-x^2}}{2} + \frac{a^2}{2}\sin^{-1}\frac{x}{a}$

14.114 $\displaystyle\int \frac{dx}{x\sqrt{a^2-x^2}} = -\frac{1}{a}\ln\left(\frac{a+\sqrt{a^2-x^2}}{x}\right)$

14.115 $\displaystyle\int \frac{dx}{x^2\sqrt{a^2-x^2}} = -\frac{\sqrt{a^2-x^2}}{a^2x}$

14.116 $\displaystyle\int \sqrt{a^2-x^2}\,dx = \frac{x\sqrt{a^2-x^2}}{2} + \frac{a^2}{2}\sin^{-1}\frac{x}{a}$

14.117 $\displaystyle\int x\sqrt{a^2-x^2}\,dx = -\frac{(a^2-x^2)^{3/2}}{3}$

14.118 $\displaystyle\int x^2\sqrt{a^2-x^2}\,dx = -\frac{x(a^2-x^2)^{3/2}}{4} + \frac{a^2x\sqrt{a^2-x^2}}{8} + \frac{a^4}{8}\sin^{-1}\frac{x}{a}$

14.119 $\displaystyle\int \frac{\sqrt{a^2-x^2}}{x}\,dx = \sqrt{a^2-x^2} - a\ln\left(\frac{a+\sqrt{a^2-x^2}}{x}\right)$

14.120 $\displaystyle\int \frac{\sqrt{a^2-x^2}}{x^2}\,dx = -\frac{\sqrt{a^2-x^2}}{x} - \sin^{-1}\frac{x}{a}$

Integrals Involving sin *ax*

14.121 $\displaystyle\int \sin ax \, dx = -\frac{\cos ax}{a}$

14.122 $\displaystyle\int x \sin ax \, dx = \frac{\sin ax}{a^2} - \frac{x \cos ax}{a}$

14.123 $\displaystyle\int x^2 \sin ax \, dx = \frac{2x}{a^2} \sin ax + \left(\frac{2}{a^3} - \frac{x^2}{a}\right) \cos ax$

14.124 $\displaystyle\int \frac{\sin ax}{x} \, dx = ax - \frac{(ax)^3}{3 \cdot 3!} + \frac{(ax)^5}{5 \cdot 5!} - \cdots$

14.125 $\displaystyle\int \frac{\sin ax}{x^2} \, dx = -\frac{\sin ax}{x} + a \int \frac{\cos ax}{x} \, dx$ [see 17.18.5]

14.126 $\displaystyle\int \frac{dx}{\sin ax} = \frac{1}{a} \ln(\csc ax - \cot ax) = \frac{1}{a} \ln \tan \frac{ax}{2}$

14.127 $\displaystyle\int \frac{x \, dx}{\sin ax} = \frac{1}{a^2} \left\{ ax + \frac{(ax)^3}{18} + \frac{7(ax)^5}{1800} + \cdots + \frac{2(2^{2n-1} - 1)B_n(ax)^{2n+1}}{(2n+1)!} + \cdots \right.$

14.128 $\displaystyle\int \sin^2 ax \, dx = \frac{x}{2} - \frac{\sin 2ax}{4a}$

14.129 $\displaystyle\int x \sin^2 ax \, dx = \frac{x^2}{4} - \frac{x \sin 2ax}{4a} - \frac{\cos 2ax}{8a^2}$

Integrals Involving cos *ax*

14.130 $\displaystyle\int \cos ax \, dx = \frac{\sin ax}{a}$

14.131 $\displaystyle\int x \cos ax \, dx = \frac{\cos ax}{a^2} + \frac{x \sin ax}{a}$

14.132 $\displaystyle\int x^2 \cos ax \, dx = \frac{2x}{a^2} \cos ax + \left(\frac{x^2}{a} - \frac{2}{a^3}\right) \sin ax$

14.133 $\displaystyle\int \frac{\cos ax}{x} \, dx = \ln x - \frac{(ax)^2}{2 \cdot 2!} + \frac{(ax)^4}{4 \cdot 4!} - \frac{(ax)^6}{6 \cdot 6!} + \cdots$

14.134 $\displaystyle\int \frac{\cos ax}{x^2}\,dx = -\frac{\cos ax}{x} - a\int \frac{\sin ax}{x}\,dx$

14.135 $\displaystyle\int \frac{dx}{\cos ax} = \frac{1}{a}\ln(\sec ax + \tan ax) = \frac{1}{a}\ln\tan\left(\frac{\pi}{4} + \frac{ax}{2}\right)$

14.136 $\displaystyle\int \frac{x\,dx}{\cos ax} = \frac{1}{a^2}\left\{\frac{(ax)^2}{2} + \frac{(ax)^4}{8} + \frac{5(ax)^6}{144} + \cdots + \frac{E_n(ax)^{2n+2}}{(2n+2)(2n)!} + \cdots\right\}$

14.137 $\displaystyle\int \cos^2 ax\,dx = \frac{x}{2} + \frac{\sin 2ax}{4a}$

14.138 $\displaystyle\int x\cos^2 ax\,dx = \frac{x^2}{4} + \frac{x\sin 2ax}{4a} + \frac{\cos 2ax}{8a^2}$

Integrals Involving sin *ax* and cos *ax*

14.139 $\displaystyle\int \sin ax\cos ax\,dx = \frac{\sin^2 ax}{2a}$

14.140 $\displaystyle\int \sin px\cos qx\,dx = -\frac{\cos(p-q)x}{2(p-q)} - \frac{\cos(p+q)x}{2(p+q)}$

14.141 $\displaystyle\int \sin^n ax\cos ax\,dx = \frac{\sin^{n+1} ax}{(n+1)a}$

14.142 $\displaystyle\int \cos^n ax\sin ax\,dx = -\frac{\cos^{n+1} ax}{(n+1)a}$

14.143 $\displaystyle\int \sin^2 ax\cos^2 ax\,dx = \frac{x}{8} - \frac{\sin 4ax}{32a}$

14.144 $\displaystyle\int \frac{dx}{\sin ax\cos ax} = \frac{1}{a}\ln\tan ax$

14.145 $\displaystyle\int \frac{dx}{\sin^2 ax\cos ax} = \frac{1}{a}\ln\tan\left(\frac{\pi}{4} + \frac{ax}{2}\right) - \frac{1}{a\sin ax}$

14.146 $\displaystyle\int \frac{dx}{\sin ax\cos^2 ax} = \frac{1}{a}\ln\tan\frac{ax}{2} + \frac{1}{a\cos ax}$

14.147 $\displaystyle\int \frac{dx}{\sin^2 ax \cos^2 ax} = -\frac{2\cot 2ax}{a}$

14.148 $\displaystyle\int \frac{\sin^2 ax}{\cos ax}\,dx = -\frac{\sin ax}{a} + \frac{1}{a}\ln\tan\left(\frac{ax}{2} + \frac{\pi}{4}\right)$

14.149 $\displaystyle\int \frac{\cos^2 ax}{\sin ax}\,dx = \frac{\cos ax}{a} + \frac{1}{a}\ln\tan\frac{ax}{2}$

Integrals Involving tan ax

14.150 $\displaystyle\int \tan ax\,dx = -\frac{1}{a}\ln\cos ax = \frac{1}{a}\ln\sec ax$

14.151 $\displaystyle\int \tan^2 ax\,dx = \frac{\tan ax}{a} - x$

14.152 $\displaystyle\int \tan^3 ax\,dx = \frac{\tan^2 ax}{2a} + \frac{1}{a}\ln\cos ax$

14.153 $\displaystyle\int \tan^n ax \sec^2 ax\,dx = \frac{\tan^{n+1} ax}{(n+1)a}$

14.154 $\displaystyle\int \frac{\sec^2 ax}{\tan ax}\,dx = \frac{1}{a}\ln\tan ax$

14.155 $\displaystyle\int \frac{dx}{\tan ax} = \frac{1}{a}\ln\sin ax$

14.156

$\displaystyle\int x\tan ax\,dx = \frac{1}{a^2}\left\{\frac{(ax)^3}{3} + \frac{(ax)^5}{15} + \frac{2(ax)^7}{105} + \cdots + \frac{2^{2n}(2^{2n}-1)B_n(ax)^{2n+1}}{(2n+1)!} + \cdots\right.$

14.157 $\displaystyle\int \frac{\tan ax}{x}\,dx = ax + \frac{(ax)^3}{9} + \frac{2(ax)^5}{75} + \cdots + \frac{2^{2n}(2^{2n}-1)B_n(ax)^{2n-1}}{(2n-1)(2n)!} + \cdots$

14.158 $\displaystyle\int x\tan^2 ax\,dx = \frac{x\tan ax}{a} + \frac{1}{a^2}\ln\cos ax - \frac{x^2}{2}$

14.159 $\displaystyle\int \frac{dx}{p + q\tan ax} = \frac{px}{p^2+q^2} + \frac{q}{a(p^2+q^2)}\ln(q\sin ax + p\cos ax)$

14.160 $\displaystyle\int \tan^n ax\, dx = \frac{\tan^{n-1} ax}{(n-1)a} - \int \tan^{n-2} ax\, dx$

Integrals Involving cot ax

14.161 $\displaystyle\int \cot ax\, dx = \frac{1}{a} \ln \sin ax$

14.162 $\displaystyle\int \cot^2 ax\, dx = -\frac{\cot ax}{a} - x$

14.163 $\displaystyle\int \cot^3 ax\, dx = -\frac{\cot^2 ax}{2a} - \frac{1}{a} \ln \sin ax$

14.164 $\displaystyle\int \cot^n ax \csc^2 ax\, dx = -\frac{\cot^{n+1} ax}{(n+1)a}$

14.165 $\displaystyle\int \frac{\csc^2 ax}{\cot ax}\, dx = -\frac{1}{a} \ln \cot ax$

14.166 $\displaystyle\int \frac{dx}{\cot ax} = -\frac{1}{a} \ln \cos ax$

14.167 $\displaystyle\int x \cot ax\, dx = \frac{1}{a^2}\left\{ ax - \frac{(ax)^3}{9} - \frac{(ax)^5}{225} - \cdots - \frac{2^{2n} B_n (ax)^{2n+1}}{(2n+1)!} - \cdots \right.$

14.168 $\displaystyle\int \frac{\cot ax}{x}\, dx = -\frac{1}{ax} - \frac{ax}{3} - \frac{(ax)^3}{135} - \cdots - \frac{2^{2n} B_n (ax)^{2n-1}}{(2n-1)(2n)!} - \cdots$

14.169 $\displaystyle\int x \cot^2 ax\, dx = -\frac{x \cot ax}{a} + \frac{1}{a^2} \ln \sin ax - \frac{x^2}{2}$

14.170 $\displaystyle\int \frac{dx}{p + q \cot ax} = \frac{px}{p^2 + q^2} - \frac{q}{a(p^2 + q^2)} \ln (q \sin ax + q \cos ax)$

14.171 $\displaystyle\int \cot^n ax\, dx = -\frac{\cot^{n-1} ax}{(n-1)a} - \int \cot^{n-2} ax\, dx$

Integrals Involving sec *ax*

14.172 $\displaystyle\int \sec ax\, dx = \frac{1}{a} \ln (\sec ax + \tan ax) = \frac{1}{a} \ln \tan \left(\frac{ax}{2} + \frac{\pi}{4} \right)$

14.173 $\displaystyle\int \sec^2 ax\, dx = \frac{\tan ax}{a}$

14.174 $\displaystyle\int \sec^3 ax\, dx = \frac{\sec ax \tan ax}{2a} + \frac{1}{2a} \ln (\sec ax + \tan ax)$

14.175 $\displaystyle\int \sec^n ax \tan ax\, dx = \frac{\sec^n ax}{na}$

14.176 $\displaystyle\int \frac{dx}{\sec ax} = \frac{\sin ax}{a}$

14.177 $\displaystyle\int x \sec ax\, dx = \frac{1}{a^2} \left\{ \frac{(ax)^2}{2} + \frac{(ax)^4}{8} + \frac{5(ax)^6}{144} + \cdots + \frac{E_n (ax)^{2n+2}}{(2n+2)(2n)!} + \cdots \right\}$

14.178 $\displaystyle\int \frac{\sec ax}{x}\, dx = \ln x + \frac{(ax)^2}{4} + \frac{5(ax)^4}{96} + \frac{61(ax)^6}{4320} + \cdots + \frac{E_n (ax)^{2n}}{2n(2n)!} + \cdots$

14.179 $\displaystyle\int x \sec^2 ax\, dx = \frac{x}{a} \tan ax + \frac{1}{a^2} \ln \cos ax$

14.180 $\displaystyle\int \frac{dx}{q + p \sec ax} = \frac{x}{q} - \frac{p}{q} \int \frac{dx}{p + q \cos ax}$

14.181 $\displaystyle\int \sec^n ax\, dx = \frac{\sec^{n-2} ax \tan ax}{a(n-1)} + \frac{n-2}{n-1} \int \sec^{n-2} ax\, dx$

Integrals Involving csc *ax*

14.182 $\displaystyle\int \csc ax\, dx = \frac{1}{a} \ln (\csc ax - \cot ax) = \frac{1}{a} \ln \tan \frac{ax}{2}$

14.183 $\displaystyle\int \csc^2 ax\, dx = -\frac{\cot ax}{a}$

14.184 $\displaystyle \int \csc^3 ax\, dx = -\frac{\csc ax \cot ax}{2a} + \frac{1}{2a} \ln \tan \frac{ax}{2}$

14.185 $\displaystyle \int \csc^n ax \cot ax\, dx = -\frac{\csc^n ax}{na}$

14.186 $\displaystyle \int \frac{dx}{\csc ax} = -\frac{\cos ax}{a}$

14.187

$$\int x \csc ax\, dx = \frac{1}{a^2} \left\{ ax + \frac{(ax)^3}{18} + \frac{7(ax)^5}{1800} + \cdots + \frac{2(2^{2n-1}-1)B_n(ax)^{2n+1}}{(2n+1)!} + \cdots \right\}$$

14.188 $\displaystyle \int \frac{\csc ax}{x}\, dx = -\frac{1}{ax} + \frac{ax}{6} + \frac{7(ax)^3}{1080} + \cdots + \frac{2(2^{2n-1}-1)B_n(ax)^{2n-1}}{(2n-1)(2n)!} + \cdots$

14.189 $\displaystyle \int x \csc^2 ax\, dx = -\frac{x \cot ax}{a} + \frac{1}{a^2} \ln \sin ax$

14.190 $\displaystyle \int \frac{dx}{q + p \csc ax} = \frac{x}{q} - \frac{p}{q} \int \frac{dx}{p + q \sin ax}$ [See 17.17.22]

14.191 $\displaystyle \int \csc^n ax\, dx = -\frac{\csc^{n-2} ax \cot ax}{a(n-1)} + \frac{n-2}{n-1} \int \csc^{n-2} ax\, dx$

Integrals Involving Inverse Trigonometric Functions

14.192 $\displaystyle \int \sin^{-1} \frac{x}{a}\, dx = x \sin^{-1} \frac{x}{a} + \sqrt{a^2 - x^2}$

14.193 $\displaystyle \int x \sin^{-1} \frac{x}{a}\, dx = \left(\frac{x^2}{2} - \frac{a^2}{4} \right) \sin^{-1} \frac{x}{a} + \frac{x\sqrt{a^2 - x^2}}{4}$

14.194 $\displaystyle \int x^2 \sin^{-1} \frac{x}{a}\, dx = \frac{x^3}{3} \sin^{-1} \frac{x}{a} + \frac{(x^2 + 2a^2)\sqrt{a^2 - x^2}}{9}$

14.195 $\displaystyle \int \cos^{-1} \frac{x}{a}\, dx = x \cos^{-1} \frac{x}{a} - \sqrt{a^2 - x^2}$

14.196 $\displaystyle \int x \cos^{-1} \frac{x}{a}\, dx = \left(\frac{x^2}{2} - \frac{a^2}{4} \right) \cos^{-1} \frac{x}{a} - \frac{x\sqrt{a^2 - x^2}}{4}$

14.197 $\displaystyle\int x^2 \cos^{-1}\frac{x}{a}\,dx = \frac{x^3}{3}\cos^{-1}\frac{x}{a} - \frac{(x^2+2a^2)\sqrt{a^2-x^2}}{9}$

14.198 $\displaystyle\int \tan^{-1}\frac{x}{a}\,dx = x\tan^{-1}\frac{x}{a} - \frac{a}{2}\ln(x^2+a^2)$

14.199 $\displaystyle\int x\tan^{-1}\frac{x}{a}\,dx = \tfrac{1}{2}(x^2+a^2)\tan^{-1}\frac{x}{a} - \frac{ax}{2}$

14.200 $\displaystyle\int x^2\tan^{-1}\frac{x}{a}\,dx = \frac{x^3}{3}\tan^{-1}\frac{x}{a} - \frac{ax^2}{6} + \frac{a^3}{6}\ln(x^2+a^2)$

14.201 $\displaystyle\int \cot^{-1}\frac{x}{a}\,dx = x\cot^{-1}\frac{x}{a} + \frac{a}{2}\ln(x^2+a^2)$

14.202 $\displaystyle\int x\cot^{-1}\frac{x}{a}\,dx = \tfrac{1}{2}(x^2+a^2)\cot^{-1}\frac{x}{a} + \frac{ax}{2}$

14.203 $\displaystyle\int x^2\cot^{-1}\frac{x}{a}\,dx = \frac{x^3}{3}\cot^{-1}\frac{x}{a} + \frac{ax^2}{6} - \frac{a^3}{6}\ln(x^2+a^2)$

14.204 $\displaystyle\int \sec^{-1}\frac{x}{a}\,dx = \begin{cases} x\sec^{-1}\dfrac{x}{a} - a\ln(x+\sqrt{x^2-a^2}) & 0 < \sec^{-1}\dfrac{x}{a} < \dfrac{\pi}{2} \\[2ex] x\sec^{-1}\dfrac{x}{a} + a\ln(x+\sqrt{x^2-a^2}) & \dfrac{\pi}{2} < \sec^{-1}\dfrac{x}{a} < \pi \end{cases}$

14.205 $\displaystyle\int x\sec^{-1}\frac{x}{a}\,dx = \begin{cases} \dfrac{x^2}{2}\sec^{-1}\dfrac{x}{a} - \dfrac{a\sqrt{x^2-a^2}}{2} & 0 < \sec^{-1}\dfrac{x}{a} < \dfrac{\pi}{2} \\[2ex] \dfrac{x^2}{2}\sec^{-1}\dfrac{x}{a} + \dfrac{a\sqrt{x^2-a^2}}{2} & \dfrac{\pi}{2} < \sec^{-1}\dfrac{x}{a} < \pi \end{cases}$

14.206 $\displaystyle\int x^2\sec^{-1}\frac{x}{a}\,dx$

$= \begin{cases} \dfrac{x^3}{3}\sec^{-1}\dfrac{x}{a} - \dfrac{ax\sqrt{x^2-a^2}}{6} - \dfrac{a^3}{6}\ln(x+\sqrt{x^2-a^2}) & 0 < \sec^{-1}\dfrac{x}{a} \\[2ex] \dfrac{x^3}{3}\sec^{-1}\dfrac{x}{a} + \dfrac{ax\sqrt{x^2-a^2}}{6} + \dfrac{a^3}{6}\ln(x+\sqrt{x^2-a^2}) & \dfrac{\pi}{2} < \sec^{-1}\dfrac{x}{a} \end{cases}$

14.207 $\displaystyle\int \csc^{-1}\frac{x}{a}\,dx = \begin{cases} x\csc^{-1}\dfrac{x}{a} + a\ln(x+\sqrt{x^2-a^2}) & 0 < \csc^{-1}\dfrac{x}{a} < \dfrac{\pi}{2} \\[3mm] x\csc^{-1}\dfrac{x}{a} - a\ln(x+\sqrt{x^2-a^2}) & -\dfrac{\pi}{2} < \csc^{-1}\dfrac{x}{a} < 0 \end{cases}$

14.208 $\displaystyle\int x\csc^{-1}\frac{x}{a}\,dx = \begin{cases} \dfrac{x^2}{2}\csc^{-1}\dfrac{x}{a} + \dfrac{a\sqrt{x^2-a^2}}{2} & 0 < \csc^{-1}\dfrac{x}{a} < \dfrac{\pi}{2} \\[3mm] \dfrac{x^2}{2}\csc^{-1}\dfrac{x}{a} - \dfrac{a\sqrt{x^2-a^2}}{2} & -\dfrac{\pi}{2} < \csc^{-1}\dfrac{x}{a} < 0 \end{cases}$

14.209

$$\int x^2\csc^{-1}\frac{x}{a}\,dx$$

$$= \begin{cases} \dfrac{x^3}{3}\csc^{-1}\dfrac{x}{a} + \dfrac{ax\sqrt{x^2-a^2}}{6} + \dfrac{a^3}{6}\ln(x+\sqrt{x^2-a^2}) & 0 < \csc^{-1}\dfrac{x}{a} < \dfrac{\pi}{2} \\[3mm] \dfrac{x^3}{3}\csc^{-1}\dfrac{x}{a} - \dfrac{ax\sqrt{x^2-a^2}}{6} - \dfrac{a^3}{6}\ln(x+\sqrt{x^2-a^2}) & -\dfrac{\pi}{2} < \csc^{-1}\dfrac{x}{a} < 0 \end{cases}$$

14.210 $\displaystyle\int x^m\sin^{-1}\frac{x}{a}\,dx = \frac{x^{m+1}}{m+1}\sin^{-1}\frac{x}{a} - \frac{1}{m+1}\int \frac{x^{m+1}}{\sqrt{a^2-x^2}}\,dx$

14.211 $\displaystyle\int x^m\cos^{-1}\frac{x}{a}\,dx = \frac{x^{m+1}}{m+1}\cos^{-1}\frac{x}{a} + \frac{1}{m+1}\int \frac{x^{m+1}}{\sqrt{a^2-x^2}}\,dx$

14.212 $\displaystyle\int x^m\tan^{-1}\frac{x}{a}\,dx = \frac{x^{m+1}}{m+1}\tan^{-1}\frac{x}{a} - \frac{a}{m+1}\int \frac{x^{m+1}}{x^2+a^2}\,dx$

14.213 $\displaystyle\int x^m\cot^{-1}\frac{x}{a}\,dx = \frac{x^{m+1}}{m+1}\cot^{-1}\frac{x}{a} + \frac{a}{m+1}\int \frac{x^{m+1}}{x^2+a^2}\,dx$

14.214 $\displaystyle\int x^m\sec^{-1}\frac{x}{a}\,dx$

$$= \begin{cases} \dfrac{x^{m+1}\sec^{-1}(x/a)}{m+1} - \dfrac{a}{m+1}\displaystyle\int \dfrac{x^m\,dx}{\sqrt{x^2-a^2}} & 0 < \sec^{-1}\dfrac{x}{a} < \dfrac{\pi}{2} \\[3mm] \dfrac{x^{m+1}\sec^{-1}(x/a)}{m+1} + \dfrac{a}{m+1}\displaystyle\int \dfrac{x^m\,dx}{\sqrt{x^2-a^2}} & \dfrac{\pi}{2} < \sec^{-1}\dfrac{x}{a} < \pi \end{cases}$$

14.215 $\int x^m \csc^{-1} \dfrac{x}{a} \, dx$

$$= \begin{cases} \dfrac{x^{m+1} \csc^{-1}(x/a)}{m+1} + \dfrac{a}{m+1} \displaystyle\int \dfrac{x^m \, dx}{\sqrt{x^2 - a^2}} & 0 < \csc^{-1}\dfrac{x}{a} < \dfrac{\pi}{2} \\[4mm] \dfrac{x^{m+1} \csc^{-1}(x/a)}{m+1} - \dfrac{a}{m+1} \displaystyle\int \dfrac{x^m \, dx}{\sqrt{x^2 - a^2}} & -\dfrac{\pi}{2} < \csc^{-1}\dfrac{x}{a} < 0 \end{cases}$$

Integrals Involving e^{ax}

14.216 $\int e^{ax} \, dx = \dfrac{e^{ax}}{a}$

14.217 $\int x e^{ax} \, dx = \dfrac{e^{ax}}{a}\left(x - \dfrac{1}{a}\right)$

14.218 $\int x^2 e^{ax} \, dx = \dfrac{e^{ax}}{a}\left(x^2 - \dfrac{2x}{a} + \dfrac{2}{a^2}\right)$

14.219 $\int x^n e^{ax} \, dx = \dfrac{x^n e^{ax}}{a} - \dfrac{n}{a}\int x^{n-1} e^{ax} \, dx$

$$= \dfrac{e^{ax}}{a}\left(x^n - \dfrac{nx^{n-1}}{a} + \dfrac{n(n-1)x^{n-2}}{a^2} - \cdots \dfrac{(-1)^n n!}{a^n}\right)$$

if n = positive integer

14.220 $\int \dfrac{e^{ax}}{x} \, dx = \ln x + \dfrac{ax}{1 \cdot 1!} + \dfrac{(ax)^2}{2 \cdot 2!} + \dfrac{(ax)^3}{3 \cdot 3!} + \cdots$

14.221 $\int \dfrac{e^{ax}}{x^n} \, dx = \dfrac{-e^{ax}}{(n-1)x^{n-1}} + \dfrac{a}{n-1}\int \dfrac{e^{ax}}{x^{n-1}} \, dx$

14.222 $\int \dfrac{dx}{p + qe^{ax}} = \dfrac{x}{p} - \dfrac{1}{ap}\ln(p + qe^{ax})$

14.223 $\displaystyle\int \frac{dx}{(p + qe^{ax})^2} = \frac{x}{p^2} + \frac{1}{ap(p + qe^{ax})} - \frac{1}{ap^2}\ln(p + qe^{ax})$

14.224 $\displaystyle\int \frac{dx}{pe^{ax} + qe^{-ax}} = \begin{cases} \dfrac{1}{a\sqrt{pq}}\tan^{-1}\left(\sqrt{\dfrac{p}{q}}\,e^{ax}\right) \\[4mm] \dfrac{1}{2a\sqrt{-pq}}\ln\left(\dfrac{e^{ax} - \sqrt{-q/p}}{e^{ax} + \sqrt{-q/p}}\right) \end{cases}$

14.225 $\displaystyle\int e^{ax}\sin bx\, dx = \frac{e^{ax}(a\sin bx - b\cos bx)}{a^2 + b^2}$

14.226 $\displaystyle\int e^{ax}\cos bx\, dx = \frac{e^{ax}(a\cos bx + b\sin bx)}{a^2 + b^2}$

14.227 $\displaystyle\int xe^{ax}\sin bx\, dx$

$\displaystyle = \frac{xe^{ax}(a\sin bx - b\cos bx)}{a^2 + b^2} - \frac{e^{ax}\{(a^2 - b^2)\sin bx - 2ab\cos bx\}}{(a^2 + b^2)^2}$

14.228 $\displaystyle\int xe^{ax}\cos bx\, dx$

$\displaystyle = \frac{xe^{ax}(a\cos bx + b\sin bx)}{a^2 + b^2} - \frac{e^{ax}\{(a^2 - b^2)\cos bx - 2ab\sin bx\}}{(a^2 + b^2)^2}$

14.229 $\displaystyle\int e^{ax}\ln x\, dx = \frac{e^{ax}\ln x}{a} - \frac{1}{a}\int \frac{e^{ax}}{x}\, dx$

14.230 $\displaystyle\int e^{ax}\sin^n bx\, dx$

$\displaystyle = \frac{e^{ax}\sin^{n-1} bx}{a^2 + n^2 b^2}(a\sin bx - nb\cos bx) + \frac{n(n-1)b^2}{a^2 + n^2 b^2}\int e^{ax}\sin^{n-2} bx\, dx$

14.231 $\displaystyle\int e^{ax}\cos^n bx\, dx$

$\displaystyle = \frac{e^{ax}\cos^{n-1} bx}{a^2 + n^2 b^2}(a\cos bx + nb\sin bx) + \frac{n(n-1)b^2}{a^2 + n^2 b^2}\int e^{ax}\cos^{n-2} bx\, dx$

Integrals Involving ln x

14.232 $\displaystyle\int \ln x\, dx = x \ln x - x$

14.233 $\displaystyle\int x \ln x\, dx = \frac{x^2}{2}\left(\ln x - \tfrac{1}{2}\right)$

14.234 $\displaystyle\int x^m \ln x\, dx = \frac{x^{m+1}}{m+1}\left(\ln x - \frac{1}{m+1}\right)$ [If $m = -1$, see 14.235]

14.235 $\displaystyle\int \frac{\ln x}{x}\, dx = \frac{1}{2}\ln^2 x$

14.236 $\displaystyle\int \frac{\ln x}{x^2}\, dx = -\frac{\ln x}{x} - \frac{1}{x}$

14.237 $\displaystyle\int \ln^2 x\, dx = x \ln^2 x - 2x \ln x + 2x$

14.238 $\displaystyle\int \frac{\ln^n x\, dx}{x} = \frac{\ln^{n+1} x}{n+1}$ [If $n = -1$, see 14.239]

14.239 $\displaystyle\int \frac{dx}{x \ln x} = \ln(\ln x)$

14.240 $\displaystyle\int \frac{dx}{\ln x} = \ln(\ln x) + \ln x + \frac{\ln^2 x}{2\cdot 2!} + \frac{\ln^3 x}{3\cdot 3!} + \cdots$

14.241 $\displaystyle\int \frac{x^m\, dx}{\ln x} = \ln(\ln x) + (m+1)\ln x + \frac{(m+1)^2 \ln^2 x}{2\cdot 2!} + \frac{(m+1)^3 \ln^3}{3\cdot 3!}$

14.242 $\displaystyle\int \ln^n x\, dx = x \ln^n x - n \int \ln^{n-1} x\, dx$

14.243 $\displaystyle\int x^m \ln^n x\, dx = \frac{x^{m+1} \ln^n x}{m+1} - \frac{n}{m+1}\int x^m \ln^{n-1} x\, dx$

[If $m = -1$, see 14.238]

14.244 $\displaystyle\int \frac{dx}{\ln x} = \ln(\ln x) + \ln x + \frac{\ln^2 x}{2 \cdot 2!} + \frac{\ln^3 x}{3 \cdot 3!} + \cdots$

14.245 $\displaystyle\int \frac{x^m \, dx}{\ln x} = \ln(\ln x) + (m+1)\ln x + \frac{(m+1)^2 \ln^2 x}{2 \cdot 2!} + \frac{(m+1)^3 \ln^3}{3 \cdot 3!}$

14.246 $\displaystyle\int \ln^n x \, dx = x \ln^n x - n \int \ln^{n-1} x \, dx$

15. Definite Integrals

Definition of a Definite Integral

Let $f(x)$ be defined in an interval $a \le x \le b$. Divide the interval into n equal parts of length $Dx = (b - a)/n$. Then the definite integral of $f(x)$ between $x = a$ and $x = b$ is defined as

15.1

$$\int_a^b f(x)\,dx =$$

$$\lim_{n\to\infty} \{f(a)\,\Delta x + f(a + \Delta x)\,\Delta x + f(a + 2\Delta x)\,\Delta x + \cdots + f(a + (n-1)\,\Delta x)\,\Delta x\}$$

The limit will certainly exist if $f(x)$ is piecewise continuous.

If $f(x) = \dfrac{d}{dx} g(x)$, then by the fundamental theorem of the integral calculus, the above definite integral can be evaluated by using the result

15.2 $\displaystyle\int_a^b f(x)\,dx = \int_a^b \frac{d}{dx} g(x)\,dx = g(x)\Big|_a^b = g(b) - g(a)$

If the interval is infinite or if $f(x)$ has a singularity at some point in the interval, the definite integral is called an *improper integral* and can be defined by using appropriate limiting procedures. For example,

15.3 $\displaystyle\int_a^\infty f(x)\,dx = \lim_{b\to\infty} \int_a^b f(x)\,dx$

15.4 $\displaystyle\int_{-\infty}^\infty f(x)\,dx = \lim_{\substack{a\to-\infty \\ b\to\infty}} \int_a^b f(x)\,dx$

15.5 $\displaystyle\int_a^b f(x)\,dx = \lim_{\epsilon\to 0} \int_a^{b-\epsilon} f(x)\,dx$

if b is a singular point.

15.6 $\displaystyle\int_a^b f(x)\,dx = \lim_{\epsilon\to 0} \int_{a+\epsilon}^b f(x)\,dx$

if a is a singular point.

General Formulas Involving Definite Integrals

15.7 $\displaystyle\int_a^b \{f(x) \pm g(x) \pm h(x) \pm \cdots\}\,dx =$

$\displaystyle\int_a^b f(x)\,dx \pm \int_a^b g(x)\,dx \pm \int_a^b h(x)\,dx \pm \cdots$

15.8 $\displaystyle\int_a^b cf(x)\,dx = c\int_a^b f(x)\,dx$ where c is any constant.

15.9 $\displaystyle\int_a^a f(x)\,dx = 0$

15.10 $\displaystyle\int_a^b f(x)\,dx = -\int_b^a f(x)\,dx$

15.11 $\displaystyle\int_a^b f(x)\,dx = \int_a^c f(x)\,dx + \int_c^b f(x)\,dx$

15.12 $\displaystyle\int_a^b f(x)\,dx = (b-a)f(c)$ where c is between a and b.

This is called the *mean value theorem* for definite integrals and is valid if $f(x)$ is continuous in $a \leq x \leq b$.

$$15.13 \qquad \int_a^b f(x) g(x) \, dx = f(c) \int_a^b g(x) \, dx$$

where c is between a and b.

This is a generalization of 15.12 and is valid if $f(x)$ and $g(x)$ are continuous in $a \leq x \leq b$ and $g(x) \geq 0$.

Leibnitz's Rule for Differentiation of Integrals

$$15.14 \qquad \frac{d}{d\alpha} \int_{\phi_1(\alpha)}^{\phi_2(\alpha)} F(x, \alpha) \, dx = \int_{\phi_1(\alpha)}^{\phi_2(\alpha)} \frac{\partial F}{\partial \alpha} \, dx + F(\phi_2, \alpha) \frac{d\phi_2}{d\alpha} - F(\phi_1, \alpha) \frac{d\phi_1}{d\alpha}$$

Approximate Formulas for Definite Integrals

In the following, the interval from $x = a$ to $x = b$ is subdivided into n equal parts by the points $a = x_0, x_1, x_2, \ldots, x_{n-1}, x_n = b$ and we let $y_0 = f(x_0), y_1 = f(x_1), y_2 = f(x_2), \ldots, y_n = f(x_n), h = (b - a)/n$.

Rectangular formula:

$$15.15 \qquad \int_a^b f(x) \, dx \approx h(y_0 + y_1 + y_2 + \cdots + y_{n-1})$$

Trapezoidal formula:

$$15.16 \qquad \int_a^b f(x) \, dx \approx \frac{h}{2}(y_0 + 2y_1 + 2y_2 + \cdots + 2y_{n-1} + y_n)$$

Simpson's formula (or parabolic formula) for n even:

$$15.17 \qquad \int_a^b f(x) \, dx \approx \frac{h}{3}(y_0 + 4y_1 + 2y_2 + 4y_3 + \cdots + 2y_{n-2} + 4y_{n-1} + y_n)$$

Definite Integrals Involving Rational or Irrational Expressions

15.18 $\displaystyle\int_0^\infty \frac{dx}{x^2 + a^2} = \frac{\pi}{2a}$

15.19 $\displaystyle\int_0^\infty \frac{x^{p-1}\,dx}{1+x} = \frac{\pi}{\sin p\pi}, \qquad 0 < p < 1$

15.20 $\displaystyle\int_0^\infty \frac{x^m\,dx}{x^n + a^n} = \frac{\pi a^{m+1-n}}{n\sin[(m+1)\,\pi/n]}, \qquad 0 < m+1 < n$

15.21 $\displaystyle\int_0^\infty \frac{x^m\,dx}{1 + 2x\cos\beta + x^2} = \frac{\pi}{\sin m\pi}\frac{\sin m\beta}{\sin\beta}$

15.22 $\displaystyle\int_0^a \frac{dx}{\sqrt{a^2 - x^2}} = \frac{\pi}{2}$

15.23 $\displaystyle\int_0^a \sqrt{a^2 - x^2}\,dx = \frac{\pi a^2}{4}$

Definite Integrals Involving Trigonometric Functions

15.24 $\displaystyle\int_0^\pi \sin mx \sin nx\,dx = \begin{cases} 0 & m, n \text{ integers and } m \neq n \\ \pi/2 & m, n \text{ integers and } m = n \end{cases}$

15.25 $\displaystyle\int_0^\pi \cos mx \cos nx\,dx = \begin{cases} 0 & m, n \text{ integers and } m \neq n \\ \pi/2 & m, n \text{ integers and } m = n \end{cases}$

15.26

$\displaystyle\int_0^\pi \sin mx \cos nx\,dx = \begin{cases} 0 & m, n \text{ integers and } m + n \text{ even} \\ 2m/(m^2 - n^2) & m, n \text{ integers and } m + n \text{ odd} \end{cases}$

15.27 $\displaystyle\int_0^{\pi/2} \sin^2 x\,dx = \int_0^{\pi/2} \cos^2 x\,dx = \frac{\pi}{4}$

15.28

$\displaystyle\int_0^{\pi/2} \sin^{2m} x\,dx = \int_0^{\pi/2} \cos^{2m} x\,dx = \frac{1\cdot 3\cdot 5\cdots 2m-1}{2\cdot 4\cdot 6\cdots 2m}\frac{\pi}{2}, \qquad m = 1, 2, \ldots$

15.29

$\displaystyle\int_0^{\pi/2} \sin^{2m+1} x\,dx = \int_0^{\pi/2} \cos^{2m+1} x\,dx = \frac{2\cdot 4\cdot 6\cdots 2m}{1\cdot 3\cdot 5\cdots 2m+1}, \qquad m = 1, 2, \ldots$

15.30 $\displaystyle\int_0^\infty \frac{\sin px}{x}\,dx = \begin{cases} \pi/2 & p>0 \\ 0 & p=0 \\ -\pi/2 & p<0 \end{cases}$

15.31 $\displaystyle\int_0^\infty \frac{\sin px \cos qx}{x}\,dx = \begin{cases} 0 & p>q>0 \\ \pi/2 & 0<p<q \\ \pi/4 & p=q>0 \end{cases}$

15.32 $\displaystyle\int_0^\infty \frac{\sin px \sin qx}{x^2}\,dx = \begin{cases} \pi p/2 & 0<p\leqq q \\ \pi q/2 & p\geqq q>0 \end{cases}$

15.33 $\displaystyle\int_0^\pi \frac{\cos mx\,dx}{1-2a\cos x+a^2} = \frac{\pi a^m}{1-a^2}, \qquad a^2<1, \qquad m=0,1,2,\ldots$

15.34 $\displaystyle\int_0^\infty \sin ax^2\,dx = \int_0^\infty \cos ax^2\,dx = \frac{1}{2}\sqrt{\frac{\pi}{2a}}$

15.35 $\displaystyle\int_0^\infty \frac{\sin x}{\sqrt{x}}\,dx = \int_0^\infty \frac{\cos x}{\sqrt{x}}\,dx = \sqrt{\frac{\pi}{2}}$

15.36 $\displaystyle\int_0^\infty \sin ax^2 \cos 2bx\,dx = \frac{1}{2}\sqrt{\frac{\pi}{2a}}\left(\cos\frac{b^2}{a}-\sin\frac{b^2}{a}\right)$

15.37 $\displaystyle\int_0^\infty \cos ax^2 \cos 2bx\,dx = \frac{1}{2}\sqrt{\frac{\pi}{2a}}\left(\cos\frac{b^2}{a}+\sin\frac{b^2}{a}\right)$

Definite Integrals Involving Exponential Functions

Some integrals contain Euler's constant $\gamma = 0.5772156$.

15.38 $\displaystyle\int_0^\infty e^{-ax}\cos bx\,dx = \frac{a}{a^2+b^2}$

15.39 $\displaystyle\int_0^\infty e^{-ax}\sin bx\,dx = \frac{b}{a^2+b^2}$

15.40 $\displaystyle\int_0^\infty \frac{e^{-ax}-e^{-bx}}{x}\,dx = \ln\frac{b}{a}$

15.41 $\displaystyle\int_0^\infty e^{-ax^2}\,dx = \frac{1}{2}\sqrt{\frac{\pi}{a}}$

15.42 $\displaystyle\int_0^\infty e^{-ax^2}\cos bx\,dx = \frac{1}{2}\sqrt{\frac{\pi}{a}}e^{-b^2/4a}$

15.43 $\displaystyle\int_0^\infty \frac{x\,dx}{e^x-1} = \frac{1}{1^2}+\frac{1}{2^2}+\frac{1}{3^2}+\frac{1}{4^2}+\cdots = \frac{\pi^2}{6}$

15.44 $\displaystyle\int_0^\infty \frac{x\,dx}{e^x+1} = \frac{1}{1^2}-\frac{1}{2^2}+\frac{1}{3^2}-\frac{1}{4^2}+\cdots = \frac{\pi^2}{12}$

15.45 $\displaystyle\int_0^\infty \left(\frac{1}{1+x}-e^{-x}\right)\frac{dx}{x} = \gamma$

15.46 $\displaystyle\int_0^\infty \frac{e^{-x^2}-e^{-x}}{x}\,dx = \tfrac{1}{2}\gamma$

15.47 $\displaystyle\int_0^\infty \left(\frac{1}{e^x-1}-\frac{e^{-x}}{x}\right)dx = \gamma$

Definite Integrals Involving Logarithmic Functions

15.48 $\displaystyle\int_0^1 x^m(\ln x)^n\,dx = \frac{(-1)^n n!}{(m+1)^{n+1}}$ $m>-1, n=0,1,2,\ldots$

15.49 $\displaystyle\int_0^1 \frac{\ln x}{1+x}\,dx = -\frac{\pi^2}{12}$

15.50 $\displaystyle\int_0^1 \frac{\ln x}{1-x}\,dx = -\frac{\pi^2}{6}$

15.51 $\displaystyle\int_0^1 \frac{\ln(1+x)}{x}\,dx = \frac{\pi^2}{12}$

15.52 $\displaystyle\int_0^1 \frac{\ln(1-x)}{x}\,dx = -\frac{\pi^2}{6}$

15.53 $\displaystyle\int_0^\infty e^{-x} \ln x\, dx = -\gamma$

15.54 $\displaystyle\int_0^\infty e^{-x^2} \ln x\, dx = -\frac{\sqrt{\pi}}{4}(\gamma + 2\ln 2)$

15.55 $\displaystyle\int_0^{\pi/2} \ln \sin x\, dx = \int_0^{\pi/2} \ln \cos x\, dx = -\frac{\pi}{2}\ln 2$

15.56 $\displaystyle\int_0^{\pi/2} (\ln \sin x)^2\, dx = \int_0^{\pi/2} (\ln \cos x)^2\, dx = \frac{\pi}{2}(\ln 2)^2 + \frac{\pi^3}{24}$

15.57

$\displaystyle\int_0^{2\pi} \ln(a + b\sin x)\, dx = \int_0^{2\pi} \ln(a + b\cos x)\, dx = 2\pi \ln(a + \sqrt{a^2 - b^2})$

15.58 $\displaystyle\int_0^\pi \ln(a + b\cos x)\, dx = \pi \ln\left(\frac{a + \sqrt{a^2 - b^2}}{2}\right)$

Section V

DIFFERENTIAL EQUATIONS

16. Basic Differential Equations and Solutions

Differential Equation	Solution
16.1 Separation of variables	
$f_1(x)\,g_1(y)\,dx + f_2(x)\,g_2(y)\,dy = 0$	$\displaystyle \int \frac{f_1(x)}{f_2(x)}\,dx + \int \frac{g_2(y)}{g_1(y)}\,dy = c$
16.2 Linear first-order equation	
$\displaystyle \frac{dy}{dx} + P(x)\,y = Q(x)$	$\displaystyle y\,e^{\int P\,dx} = \int Q\,e^{\int P\,dx}\,dx + c$
16.3 Bernoulli's equation	
$\displaystyle \frac{dy}{dx} + P(x)\,y = Q(x)\,y^n$	$\displaystyle v\,e^{(1-n)\int P\,dx} = (1-n)\int Q\,e^{(1-n)\int P\,dx}\,dx +$ where $v = y^{1-n}$. If $n = 1$, the solution $\displaystyle \ln y = \int (Q - P)\,dx + c$

16.4 Exact equation

$$M(x, y)\,dx + N(x, y)\,dy = 0$$

where $\partial M/\partial y = \partial N/\partial x$.

$$\int M\,\partial x + \int \left(N - \frac{\partial}{\partial y} \int M\,\partial x\right) dy = c$$

where ∂x indicates that the integration is to be performed with respect to x keeping y constant.

16.5 Homogeneous equation

$$\frac{dy}{dx} = F\left(\frac{y}{x}\right)$$

$$\ln x = \int \frac{dv}{F(v) - v} + c$$

where $v = y/x$. If $F(v) = v$, the solution is $y = cx$.

16.6 Linear, homogeneous second-order equation

$$\frac{d^2 y}{dx^2} + a\frac{dy}{dx} + by = 0$$

a, b are real constants.

Let m_1, m_2 be the roots of $m^2 + am + b = 0$. Then there are 3 cases.

Case 1. m_1, m_2 real and distinct:
$$y = c_1 e^{m_1 x} + c_2 e^{m_2 x}$$

Case 2. m_1, m_2 real and equal:
$$y = c_1 e^{m_1 x} + c_2 x e^{m_1 x}$$

Case 3. $m_1 = p + qi$, $m_2 = p - qi$:
$$y = e^{px}(c_1 \cos qx + c_2 \sin qx)$$
where $p = -a/2$, $q = \sqrt{b - a^2/4}$.

16.7 Linear, nonhomogeneous second-order equation

$$\frac{d^2 y}{dx^2} + a\frac{dy}{dx} + by = R(x)$$

a, b are real constants.

There are 3 cases corresponding to those of entry 16.7 above.

Case 1.

$$y = c_1 e^{m_1 x} + c_2 e^{m_2 x}$$

$$+ \frac{e^{m_1 x}}{m_1 - m_2} \int e^{-m_1 x} R(x)\, dx$$

$$+ \frac{e^{m_2 x}}{m_2 - m_1} \int e^{-m_2 x} R(x)\, dx$$

Case 2.

$$y = c_1 e^{m_1 x} + c_2 x e^{m_1 x}$$

$$+ x e^{m_1 x} \int e^{-m_1 x} R(x)\, dx$$

$$- e^{m_1 x} \int x e^{-m_1 x} R(x)\, dx$$

Case 3.

$$y = e^{px}(c_1 \cos qx + c_2 \sin qx)$$

$$+ \frac{e^{px} \sin qx}{q} \int e^{-px} R(x) \cos qx\, dx$$

$$- \frac{e^{px} \cos qx}{q} \int e^{-px} R(x) \sin qx\, dx$$

Section VI
SERIES

17. Series of Constants

Arithmetic Series

17.1
$$a + (a + d) + (a + 2d) + \cdots + \{a + (n - 1)d\}$$
$$= \tfrac{1}{2}n\{2a + (n - 1)d\} = \tfrac{1}{2}n(a + l)$$

where $l = a + (n \text{-} 1)d$ is the last term.

Some special cases are

17.2 $1 + 2 + 3 + \cdots + n = \tfrac{1}{2}n(n + 1)$

17.3 $1 + 3 + 5 + \cdots + (2n - 1) = n^2$

Geometric Series

17.4 $a + ar + ar^2 + ar^3 + \cdots + ar^{n-1} = \dfrac{a(1 - r^n)}{1 - r} = \dfrac{a - rl}{1 - r}$

where $l = ar^{n-1}$ is the last term and $r \neq 1$.

If $-1 < r < 1$, then

17.5 $a + ar + ar^2 + ar^3 + \cdots = \dfrac{a}{1 - r}$

113

Arithmetic-Geometric Series

17.6 $\quad a + (a+d)r + (a+2d)r^2 + \cdots + \{a+(n-1)d\}r^{n-1}$

$$= \frac{a(1-r^n)}{1-r} + \frac{rd\{1 - nr^{n-1} + (n-1)r^n\}}{(1-r)^2}$$

where $r \neq 1$.

If $-1 < r < 1$, then

17.7 $\quad a + (a+d)r + (a+2d)r^2 + \cdots = \dfrac{a}{1-r} + \dfrac{rd}{(1-r)^2}$

Sums of Powers of Positive Integers

17.8 $\quad 1^p + 2^p + 3^p + \cdots + n^p$

$$= \frac{n^{p+1}}{p+1} + \tfrac{1}{2}n^p + \frac{B_1 p n^{p-1}}{2!} - \frac{B_2 p(p-1)(p-2)n^{p-3}}{4!} + \cdots$$

where the series terminates at n^2 or n according as p is odd or even, and B_k are the *Bernoulli numbers* [see Subsection 19].

Some special cases are:

17.9 $\quad 1 + 2 + 3 + \cdots + n = \dfrac{n(n+1)}{2}$

17.10 $\quad 1^2 + 2^2 + 3^2 + \cdots + n^2 = \dfrac{n(n+1)(2n+1)}{6}$

17.11 $\quad 1^3 + 2^3 + 3^3 + \cdots + n^3 = \dfrac{n^2(n+1)^2}{4} = (1 + 2 + 3 + \cdots + n)^2$

Series Involving Reciprocals of Powers of Positive Integers

17.12 $1 - \dfrac{1}{2} + \dfrac{1}{3} - \dfrac{1}{4} + \dfrac{1}{5} - \cdots = \ln 2$

17.13 $1 - \dfrac{1}{3} + \dfrac{1}{5} - \dfrac{1}{7} + \dfrac{1}{9} - \cdots = \dfrac{\pi}{4}$

17.14 $\dfrac{1}{1^2} + \dfrac{1}{2^2} + \dfrac{1}{3^2} + \dfrac{1}{4^2} + \cdots = \dfrac{\pi^2}{6}$

17.15 $\dfrac{1}{1^4} + \dfrac{1}{2^4} + \dfrac{1}{3^4} + \dfrac{1}{4^4} + \cdots = \dfrac{\pi^4}{90}$

17.16 $\dfrac{1}{1^2} - \dfrac{1}{2^2} + \dfrac{1}{3^2} - \dfrac{1}{4^2} + \cdots = \dfrac{\pi^2}{12}$

17.17 $\dfrac{1}{1^4} - \dfrac{1}{2^4} + \dfrac{1}{3^4} - \dfrac{1}{4^4} + \cdots = \dfrac{7\pi^4}{720}$

17.18 $\dfrac{1}{1^2} + \dfrac{1}{3^2} + \dfrac{1}{5^2} + \dfrac{1}{7^2} + \cdots = \dfrac{\pi^2}{8}$

17.19 $\dfrac{1}{1^4} + \dfrac{1}{3^4} + \dfrac{1}{5^4} + \dfrac{1}{7^4} + \cdots = \dfrac{\pi^4}{96}$

17.20 $\dfrac{1}{1 \cdot 3} + \dfrac{1}{3 \cdot 5} + \dfrac{1}{5 \cdot 7} + \dfrac{1}{7 \cdot 9} + \cdots = \dfrac{1}{2}$

17.21 $\dfrac{1}{1 \cdot 3} + \dfrac{1}{2 \cdot 4} + \dfrac{1}{3 \cdot 5} + \dfrac{1}{4 \cdot 6} + \cdots = \dfrac{3}{4}$

17.22 $\dfrac{1}{1^{2p}} + \dfrac{1}{2^{2p}} + \dfrac{1}{3^{2p}} + \dfrac{1}{4^{2p}} + \cdots = \dfrac{2^{2p-1}\pi^{2p}B_p}{(2p)!}$

17.23 $\dfrac{1}{1^{2p}} + \dfrac{1}{3^{2p}} + \dfrac{1}{5^{2p}} + \dfrac{1}{7^{2p}} + \cdots = \dfrac{(2^{2p} - 1)\pi^{2p}B_p}{2(2p)!}$

17.24 $\dfrac{1}{1^{2p}} - \dfrac{1}{2^{2p}} + \dfrac{1}{3^{2p}} - \dfrac{1}{4^{2p}} + \cdots = \dfrac{(2^{2p-1} - 1)\pi^{2p}B_p}{(2p)!}$

17.25 $\dfrac{1}{1^{2p+1}} - \dfrac{1}{3^{2p+1}} + \dfrac{1}{5^{2p+1}} - \dfrac{1}{7^{2p+1}} + \cdots = \dfrac{\pi^{2p+1}E_p}{2^{2p+2}(2p)!}$

18. Binomial Series

Binomial Series

18.1 $(a+x)^n = a^n + na^{n-1}x + \dfrac{n(n-1)}{2!}a^{n-2}x^2 + \dfrac{n(n-1)(n-2)}{3!}a^{n-3}x^3 + \cdots$

$$= a^n + \binom{n}{1}a^{n-1}x + \binom{n}{2}a^{n-2}x^2 + \binom{n}{3}a^{n-3}x^3 + \cdots$$

18.2 $(a+x)^2 = a^2 + 2ax + x^2$

18.3 $(a+x)^3 = a^3 + 3a^2x + 3ax^2 + x^3$

18.4 $(a+x)^4 = a^4 + 4a^3x + 6a^2x^2 + 4ax^3 + x^4$

18.5 $(1+x)^{-1} = 1 - x + x^2 - x^3 + x^4 - \cdots$ $\qquad\qquad -1 < x < 1$

18.6 $(1+x)^{-2} = 1 - 2x + 3x^2 - 4x^3 + 5x^4 - \cdots$ $\qquad\qquad -1 < x < 1$

18.7 $(1+x)^{-3} = 1 - 3x + 6x^2 - 10x^3 + 15x^4 - \cdots$ $\qquad\qquad -1 < x < 1$

18.8 $(1+x)^{-1/2} = 1 - \dfrac{1}{2}x + \dfrac{1\cdot3}{2\cdot4}x^2 - \dfrac{1\cdot3\cdot5}{2\cdot4\cdot6}x^3 + \cdots$ $\qquad -1 < x \leq 1$

18.9 $(1+x)^{1/2} = 1 + \dfrac{1}{2}x - \dfrac{1}{2\cdot4}x^2 + \dfrac{1\cdot3}{2\cdot4\cdot6}x^3 - \cdots$ $\qquad -1 < x \leq 1$

18.10 $(1+x)^{-1/3} = 1 - \dfrac{1}{3}x + \dfrac{1\cdot4}{3\cdot6}x^2 - \dfrac{1\cdot4\cdot7}{3\cdot6\cdot9}x^3 + \cdots$ $\qquad -1 < x \leq 1$

18.11 $(1+x)^{1/3} = 1 + \dfrac{1}{3}x - \dfrac{2}{3\cdot6}x^2 + \dfrac{2\cdot5}{3\cdot6\cdot9}x^3 - \cdots$ $\qquad -1 < x \leq 1$

Series for Exponential and Logarithmic Functions

18.12 $e^x = 1 + x + \dfrac{x^2}{2!} + \dfrac{x^3}{3!} + \cdots$ $\qquad\qquad\qquad -\infty < x < \infty$

18.13 $a^x = e^{x \ln a} = 1 + x \ln a + \dfrac{(x \ln a)^2}{2!} + \dfrac{(x \ln a)^3}{3!} + \cdots$ $\qquad -\infty < x < \infty$

18.14 $\ln(1+x) = x - \dfrac{x^2}{2} + \dfrac{x^3}{3} - \dfrac{x^4}{4} + \cdots$ $-1 < x \leq 1$

18.15 $\dfrac{1}{2} \ln\left(\dfrac{1+x}{1-x}\right) = x + \dfrac{x^3}{3} + \dfrac{x^5}{5} + \dfrac{x^7}{7} + \cdots$ $-1 < x < 1$

18.16 $\ln x = 2\left\{\left(\dfrac{x-1}{x+1}\right) + \dfrac{1}{3}\left(\dfrac{x-1}{x+1}\right)^3 + \dfrac{1}{5}\left(\dfrac{x-1}{x+1}\right)^5 + \cdots\right\}$ $x > 0$

18.17 $\ln x = \left(\dfrac{x-1}{x}\right) + \dfrac{1}{2}\left(\dfrac{x-1}{x}\right)^2 + \dfrac{1}{3}\left(\dfrac{x-1}{x}\right)^3 + \cdots$ $x \geq \frac{1}{2}$

Series for Trigonometric Functions

18.18 $\sin x = x - \dfrac{x^3}{3!} + \dfrac{x^5}{5!} - \dfrac{x^7}{7!} + \cdots$ $-\infty < x < \infty$

18.19 $\cos x = 1 - \dfrac{x^2}{2!} + \dfrac{x^4}{4!} - \dfrac{x^6}{6!} + \cdots$ $-\infty < x < \infty$

18.20

$\tan x = x + \dfrac{x^3}{3} + \dfrac{2x^5}{15} + \dfrac{17x^7}{315} + \cdots + \dfrac{2^{2n}(2^{2n}-1)B_n x^{2n-1}}{(2n)!} + \cdots$ $|x| < \dfrac{\pi}{2}$

19. Bernoulli Numbers

Definition of Bernoulli Numbers

The *Bernoulli numbers* B_1, B_2, B_3, ... are defined by the series

19.1 $\dfrac{x}{e^x - 1} = 1 - \dfrac{x}{2} + \dfrac{B_1 x^2}{2!} - \dfrac{B_2 x^4}{4!} + \dfrac{B_3 x^6}{6!} - \cdots$ $|x| < 2\pi$

19.2 $1 - \dfrac{x}{2}\cot\dfrac{x}{2} = \dfrac{B_1 x^2}{2!} + \dfrac{B_3 x^4}{4!} + \dfrac{B_3 x^6}{6!} + \cdots$ $|x| < \pi$

Table of First Few Bernoulli Numbers

Bernoulli numbers
$B_1 = 1/6$
$B_2 = 1/30$
$B_3 = 1/42$
$B_4 = 1/30$
$B_5 = 5/66$
$B_6 = 691/2730$
$B_7 = 7/6$
$B_8 = 3617/510$
$B_9 = 43,867/798$
$B_{10} = 174,611/330$
$B_{11} = 854,513/138$
$B_{12} = 236,364,091/2730$

Relationship of Bernoulli Numbers

19.3

$$\binom{2n+1}{2}2^2B_1 - \binom{2n+1}{4}2^4B_2 + \binom{2n+1}{6}2^6B_3 - \cdots(-1)^{n-1}(2n+1)2^{2n}B_n = 2n$$

Series Involving Bernoulli Numbers

19.4 $\quad B_n = \dfrac{(2n)!}{2^{2n-1}\pi^{2n}}\left\{1 + \dfrac{1}{2^{2n}} + \dfrac{1}{3^{2n}} + \cdots\right\}$

19.5 $\quad B_n = \dfrac{2(2n)!}{(2^{2n}-1)\pi^{2n}}\left\{1 + \dfrac{1}{3^{2n}} + \dfrac{1}{5^{2n}} + \cdots\right\}$

19.6 $\quad B_n = \dfrac{2(2n)!}{(2^{2n-1}-1)\pi^{2n}}\left\{1 - \dfrac{1}{2^{2n}} + \dfrac{1}{3^{2n}} - \cdots\right\}$

Asymptotic Formula for Bernoulli Numbers

19.7 $\quad B_n \sim 4n^{2n}(\pi e)^{-2n}\sqrt{\pi n}$

Section VII

VECTOR ANALYSIS

20. Formulas from Vector Analysis

Vectors and Scalars

Various quantities in physics such as temperature, volume, and speed can be specified by a real number. Such quantities are called *scalars*.

Other quantities such as force, velocity, and momentum require for their specification a direction as well as magnitude. Such quantities are called *vectors*. A vector is represented by an arrow or directed line segment indicating direction. The magnitude of the vector is determined by the length of the arrow, using an appropriate unit.

Notation for Vectors

A vector is denoted by a bold-faced letter such as **A** [Fig 20-1]. The magnitude is denoted by $|\mathbf{A}|$ or A. The tail end of the arrow is called the *initial point* while the head is called the *terminal point*.

Fundamental Definitions

1. Equality of vectors. Two vectors are equal if they have the same magnitude and direction. Thus **A** = **B** in Fig. 20-1.

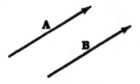

Figure 20-1

120

2. Multiplication of a vector by a scalar. If m is any real number (scalar), then mA is a vector whose magnitude is $|m|$ times the magnitude of A and whose direction is the same as or opposite to A according as $m > 0$ or $m < 0$. If $m = 0$, then mA is called the *zero* or *null vector*.

3. Sums of vectors. The sum or resultant of A and B (Fig. 20-2(a)) is a vector C = A + B formed by placing the initial point of B on the terminal point of A and joining the initial point of A to the terminal point of B as in Fig. 20-2(b). This definition is equivalent to the parallelogram law for vector addition as indicated in Fig. 20-2(c).

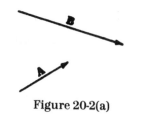

Figure 20-2(a)

Figure 20-2(b)

Figure 20-2(c)

Extensions to sums of more than two vectors are immediate. Thus, Fig. 20-3 shows how to obtain the sum E of the vectors A, B, C, and D.

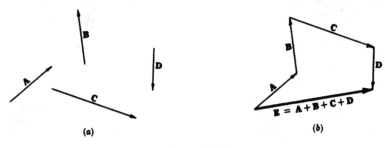

Figure 20-3

4.   Unit vectors. A *unit vector* is a vector with unit magnitude. If A is a vector, then a unit vector in the direction of A is a = A/A where $A > 0$.

Laws of Vector Algebra

If A, B, C are vectors and m, n are scalars, then:

Commutative Law for Addition

20.1   **A + B = B + A**

Associative Law for Addition

20.2   **A + (B + C) = (A + B) + C**

Associative Law for Scalar Multiplication

20.3 $m(n\mathbf{A}) = (mn)\mathbf{A} = n(m\mathbf{A})$

Distributive Law

20.4 $(m + n)\mathbf{A} = m\mathbf{A} + n\mathbf{A}$

Distributive Law

20.5 $m(\mathbf{A} + \mathbf{B}) = m\mathbf{A} + m\mathbf{B}$

Components of a Vector

A vector **A** can be represented with initial point at the origin of a rectangular coordinate system, as shown in Fig. 20-4.

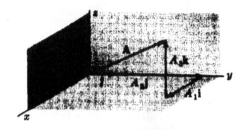

Figure 20-4

If **i**, **j**, **k** are unit vectors in the directions of the positive x, y, z axes, then:

20.6 $\mathbf{A} = A_1\mathbf{i} + A_2\mathbf{j} + A_3\mathbf{k}$

where $A_1\mathbf{i}$, $A_2\mathbf{j}$, $A_3\mathbf{k}$ are called *component vectors* of **A** in the **i**, **j**, **k** directions and A_1, A_2, A_3 are called the *components* of **A**.

Dot or Scalar Product

20.7 $\mathbf{A} \cdot \mathbf{B} = AB\cos\theta$ $\qquad 0 \leq \theta \leq \pi$

where θ is the angle between **A** and **B**.
   Fundamental results are:

Commutative Law

20.8 $\mathbf{A} \cdot \mathbf{B} = \mathbf{B} \cdot \mathbf{A}$

Distributive Law

20.9 $\mathbf{A} \cdot (\mathbf{B} + \mathbf{C}) = \mathbf{A} \cdot \mathbf{B} + \mathbf{A} \cdot \mathbf{C}$

20.10 $\mathbf{A} \cdot \mathbf{B} = A_1B_1 + A_2B_2 + A_3B_3$

where $\mathbf{A} = A_1\mathbf{i} + A_2\mathbf{j} + A_3\mathbf{k}, B = B_1\mathbf{i} + B_2\mathbf{j} + B_3\mathbf{k}.$

Cross or Vector Product

20.11 $\mathbf{A} \times \mathbf{B} = AB\sin\theta\,\mathbf{u}$ $0 \leq \theta \leq \pi$

where θ is the angle between A and B and u is a unit vector perpendicular to the plane of A and B such that A, B, u form a *right-handed system* [i.e., a right-threaded screw rotated through an angle less than 180° from A to B will advance in the direction of u as in Fig. 20-5].

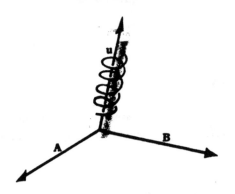

Figure 20-5

Fundamental results follow:

20.12 $\mathbf{A} \times \mathbf{B} = \begin{vmatrix} \mathbf{i} & \mathbf{j} & \mathbf{k} \\ A_1 & A_2 & A_3 \\ B_1 & B_2 & B_3 \end{vmatrix}$

$= (A_2B_3 - A_3B_2)\mathbf{i} + (A_3B_1 - A_1B_3)\mathbf{j} + (A_1B_2 - A_2B_1)\mathbf{k}$

20.13 $\mathbf{A} \times \mathbf{B} = -\mathbf{B} \times \mathbf{A}$

20.14 $\mathbf{A} \times (\mathbf{B} + \mathbf{C}) = \mathbf{A} \times \mathbf{B} + \mathbf{A} \times \mathbf{C}$

20.15 $|\mathbf{A} \times \mathbf{B}|$ = area of parallelogram having sides $\mathbf{A}$ and $\mathbf{B}$

Miscellaneous Formulas Involving Dot and Cross Products

20.16 $\mathbf{A} \cdot (\mathbf{B} \times \mathbf{C}) = \begin{vmatrix} A_1 & A_2 & A_3 \\ B_1 & B_2 & B_3 \\ C_1 & C_2 & C_3 \end{vmatrix}$

$$= A_1 B_2 C_3 + A_2 B_3 C_1 + A_3 B_1 C_2 - A_3 B_2 C_1 - A_2 B_1 C_3 - A_1 B_3 C_2$$

20.17 $|\mathbf{A} \cdot (\mathbf{B} \times \mathbf{C})|$ = volume of parallelepiped with sides $\mathbf{A}$, $\mathbf{B}$, $\mathbf{C}$

20.18 $\mathbf{A} \times (\mathbf{B} \times \mathbf{C}) = \mathbf{B}(\mathbf{A} \cdot \mathbf{C}) - \mathbf{C}(\mathbf{A} \cdot \mathbf{B})$

20.19 $(\mathbf{A} \times \mathbf{B}) \times \mathbf{C} = \mathbf{B}(\mathbf{A} \cdot \mathbf{C}) - \mathbf{A}(\mathbf{B} \cdot \mathbf{C})$

20.20 $(\mathbf{A} \times \mathbf{B}) \cdot (\mathbf{C} \times \mathbf{D}) = (\mathbf{A} \cdot \mathbf{C})(\mathbf{B} \cdot \mathbf{D}) - (\mathbf{A} \cdot \mathbf{D})(\mathbf{B} \cdot \mathbf{C})$

20.21 $(\mathbf{A} \times \mathbf{B}) \times (\mathbf{C} \times \mathbf{D}) = \mathbf{C}\{\mathbf{A} \cdot (\mathbf{B} \times \mathbf{D})\} - \mathbf{D}\{\mathbf{A} \cdot (\mathbf{B} \times \mathbf{C})\}$

PART B

TABLES

Table I Factorial n

n	$n!$
0	1 (by definition)
1	1
2	2
3	6
4	24
5	120
6	720
7	5040
8	40,320
9	362,880
10	3,628,800
11	39,916,800
12	479,001,600
13	6,227,020,800
14	87,178,291,200
15	1,307,674,368,000
16	20,922,789,888,000
17	355,687,428,096,000
18	6,402,373,705,728,000
19	121,645,100,408,832,000

n	$n!$
40	8.15915×10^{47}
41	3.34525×10^{49}
42	1.40501×10^{51}
43	6.04153×10^{52}
44	2.65827×10^{54}
45	1.19622×10^{56}
46	5.50262×10^{57}
47	2.58623×10^{59}
48	1.24139×10^{61}
49	6.08282×10^{62}
50	3.04141×10^{64}
51	1.55112×10^{66}
52	8.06582×10^{67}
53	4.27488×10^{69}
54	2.30844×10^{71}
55	1.26964×10^{73}
56	7.10999×10^{74}
57	4.05269×10^{76}
58	2.35056×10^{78}
59	1.38683×10^{80}

n	$n!$
80	7.15695×10^{118}
81	5.79713×10^{120}
82	4.75364×10^{122}
83	3.94552×10^{124}
84	3.31424×10^{126}
85	2.81710×10^{128}
86	2.42271×10^{130}
87	2.10776×10^{132}
88	1.85483×10^{134}
89	1.65080×10^{136}
90	1.48572×10^{138}
91	1.35200×10^{140}
92	1.24384×10^{142}
93	1.15677×10^{144}
94	1.08737×10^{146}
95	1.03300×10^{148}
96	9.91678×10^{149}
97	9.61928×10^{151}
98	9.42689×10^{153}
99	9.33262×10^{155}

100	9.33262×10^{157}

60	8.32099×10^{81}
61	5.07580×10^{83}
62	3.14700×10^{85}
63	1.98261×10^{87}
64	1.26887×10^{89}
65	8.24765×10^{90}
66	5.44345×10^{92}
67	3.64711×10^{94}
68	2.48004×10^{96}
69	1.71122×10^{98}
70	1.19786×10^{100}
71	8.50479×10^{101}
72	6.12345×10^{103}
73	4.47012×10^{105}
74	3.30789×10^{107}
75	2.48091×10^{109}
76	1.88549×10^{111}
77	1.45183×10^{113}
78	1.13243×10^{115}
79	8.94618×10^{116}

20	2,432,902,008,176,640,000
21	51,090,942,171,709,440,000
22	1,124,000,727,777,607,680,000
23	25,852,016,738,884,976,640,000
24	620,448,401,733,239,439,360,000
25	15,511,210,043,330,985,984,000,000
26	403,291,461,126,605,635,584,000,000
27	10,888,869,450,418,352,160,768,000,000
28	304,888,344,611,713,860,501,504,000,000
29	8,841,761,993,739,701,954,543,616,000,000
30	265,252,859,812,191,058,636,308,480,000,000
31	8.22284×10^{33}
32	2.63131×10^{35}
33	8.68332×10^{36}
34	2.95233×10^{38}
35	1.03331×10^{40}
36	3.71993×10^{41}
37	1.37638×10^{43}
38	5.23023×10^{44}
39	2.03979×10^{46}

Table II Conversion of Radians to
Degrees, Minutes, and Seconds

Radians	Deg.	Min.	Sec.	Fractions of Degrees
1	57°	17′	44.8″	57.2958°
2	114°	35′	29.6″	114.5916°
3	171°	53′	14.4″	171.8873°
4	229°	10′	59.2″	229.1831°
5	286°	28′	44.0″	286.4789°
6	343°	46′	28.8″	343.7747°
7	401°	4′	13.6″	401.0705°
8	458°	21′	58.4″	458.3662°
9	515°	39′	43.3″	515.6620°
10	572°	57′	28.1″	572.9578°
.1	5°	43′	46.5″	
.2	11°	27′	33.0″	
.3	17°	11′	19.4″	
.4	22°	55′	5.9″	
.5	28°	38′	52.4″	
.6	34°	22′	38.9″	
.7	40°	6′	25.4″	
.8	45°	50′	11.8″	
.9	51°	33′	58.3″	
.01	0°	34′	22.6″	
.02	1°	8′	45.3″	
.03	1°	43′	7.9″	
.04	2°	17′	30.6″	
.05	2°	51′	53.2″	
.06	3°	26′	15.9″	
.07	4°	0′	38.5″	
.08	4°	35′	1.2″	
.09	5°	9′	23.8″	
.001	0°	3′	26.3″	
.002	0°	6′	52.5″	
.003	0°	10′	18.8″	
.004	0°	13′	45.1″	
.005	0°	17′	11.3″	
.006	0°	20′	37.6″	
.007	0°	24′	3.9″	
.008	0°	27′	30.1″	
.009	0°	30′	56.4″	
.0001	0°	0′	20.6″	
.0002	0°	0′	41.3″	
.0003	0°	1′	1.9″	
.0004	0°	1′	22.5″	
.0005	0°	1′	43.1″	
.0006	0°	2′	3.8″	
.0007	0°	2′	24.4″	
.0008	0°	2′	45.0″	
.0009	0°	3′	5.6″	

Table III Conversion of Degrees, Minutes, and Seconds 131

Table III Conversion of Degrees, Minutes, and Seconds to Radians

Degrees	Radians
1°	.0174533
2°	.0349066
3°	.0523599
4°	.0698132
5°	.0872665
6°	.1047198
7°	.1221730
8°	.1396263
9°	.1570796
10°	.1745329

Minutes	Radians
1′	.00029089
2′	.00058178
3′	.00087266
4′	.00116355
5′	.00145444
6′	.00174533
7′	.00203622
8′	.00232711
9′	.00261800
10′	.00290888

Seconds	Radians
1″	.0000048481
2″	.0000096963
3″	.0000145444
4″	.0000193925
5″	.0000242407
6″	.0000290888
7″	.0000339370
8″	.0000387851
9″	.0000436332
10″	.0000484814

Table IV Sin x (x in degrees and minutes)

x	0′	10′	20′	30′	40′	50′
0°	.0000	.0029	.0058	.0087	.0116	.0145
1	.0175	.0204	.0233	.0262	.0291	.0320
2	.0349	.0378	.0407	.0436	.0465	.0494
3	.0523	.0552	.0581	.0610	.0640	.0669
4	.0698	.0727	.0756	.0785	.0814	.0843
5°	.0872	.0901	.0929	.0958	.0987	.1016
6	.1045	.1074	.1103	.1132	.1161	.1190
7	.1219	.1248	.1276	.1305	.1334	.1363
8	.1392	.1421	.1449	.1478	.1507	.1536
9	.1564	.1593	.1622	.1650	.1679	.1708
10°	.1736	.1765	.1794	.1822	.1851	.1880
11	.1908	.1937	.1965	.1994	.2022	.2051
12	.2079	.2108	.2136	.2164	.2193	.2221
13	.2250	.2278	.2306	.2334	.2363	.2391
14	.2419	.2447	.2476	.2504	.2532	.2560
15°	.2588	.2616	.2644	.2672	.2700	.2728
16	.2756	.2784	.2812	.2840	.2868	.2896
17	.2924	.2952	.2979	.3007	.3035	.3062
18	.3090	.3118	.3145	.3173	.3201	.3228
19	.3256	.3283	.3311	.3338	.3365	.3393
20°	.3420	.3448	.3475	.3502	.3529	.3557

x	0′	10′	20′	30′	40′	50′
45°	.7071	.7092	.7112	.7133	.7153	.7173
46	.7193	.7214	.7234	.7254	.7274	.7294
47	.7314	.7333	.7353	.7373	.7392	.7412
48	.7431	.7451	.7470	.7490	.7509	.7528
49	.7547	.7566	.7585	.7604	.7623	.7642
50°	.7660	.7679	.7698	.7716	.7735	.7753
51	.7771	.7790	.7808	.7826	.7844	.7862
52	.7880	.7898	.7916	.7934	.7951	.7969
53	.7986	.8004	.8021	.8039	.8056	.8073
54	.8090	.8107	.8124	.8141	.8158	.8175
55°	.8192	.8208	.8225	.8241	.8258	.8274
56	.8290	.8307	.8323	.8339	.8355	.8371
57	.8387	.8403	.8418	.8434	.8450	.8465
58	.8480	.8496	.8511	.8526	.8542	.8557
59	.8572	.8587	.8601	.8616	.8631	.8646
60°	.8660	.8675	.8689	.8704	.8718	.8732
61	.8746	.8760	.8774	.8788	.8802	.8816
62	.8829	.8843	.8857	.8870	.8884	.8897
63	.8910	.8923	.8936	.8949	.8962	.8975
64	.8988	.9001	.9013	.9026	.9038	.9051
65°	.9063	.9075	.9088	.9100	.9112	.9124

TABLE IV Sin x (x in degrees and minutes) 133

	0′	10′	20′	30′	40′	50′
21	.3584	.3611	.3638	.3665	.3692	.3719
22	.3746	.3773	.3800	.3827	.3854	.3881
23	.3907	.3934	.3961	.3987	.4014	.4041
24	.4067	.4094	.4120	.4147	.4173	.4200
25°	.4226	.4253	.4279	.4305	.4331	.4358
26	.4384	.4410	.4436	.4462	.4488	.4514
27	.4540	.4566	.4592	.4617	.4643	.4669
28	.4695	.4720	.4746	.4772	.4797	.4823
29	.4848	.4874	.4899	.4924	.4950	.4975
30°	.5000	.5025	.5050	.5075	.5100	.5125
31	.5150	.5175	.5200	.5225	.5250	.5275
32	.5299	.5324	.5348	.5373	.5398	.5422
33	.5446	.5471	.5495	.5519	.5544	.5568
34	.5592	.5616	.5640	.5664	.5688	.5712
35°	.5736	.5760	.5783	.5807	.5831	.5854
36	.5878	.5901	.5925	.5948	.5972	.5995
37	.6018	.6041	.6065	.6088	.6111	.6134
38	.6157	.6180	.6202	.6225	.6248	.6271
39	.6293	.6316	.6338	.6361	.6383	.6406
40°	.6428	.6450	.6472	.6494	.6517	.6539
41	.6561	.6583	.6604	.6626	.6648	.6670
42	.6691	.6713	.6734	.6756	.6777	.6799
43	.6820	.6841	.6862	.6884	.6905	.6926
44	.6947	.6967	.6988	.7009	.7030	.7050
45°	.7071	.7092	.7112	.7133	.7153	.7173

	0′	10′	20′	30′	40′	50′
66	.9135	.9147	.9159	.9171	.9182	.9194
67	.9205	.9216	.9228	.9239	.9250	.9261
68	.9272	.9283	.9293	.9304	.9315	.9325
69	.9336	.9346	.9356	.9367	.9377	.9387
70°	.9397	.9407	.9417	.9426	.9436	.9446
71	.9455	.9465	.9474	.9483	.9492	.9502
72	.9511	.9520	.9528	.9537	.9546	.9555
73	.9563	.9572	.9580	.9588	.9596	.9605
74	.9613	.9621	.9628	.9636	.9644	.9652
75°	.9659	.9667	.9674	.9681	.9689	.9696
76	.9703	.9710	.9717	.9724	.9730	.9737
77	.9744	.9750	.9757	.9763	.9769	.9775
78	.9781	.9787	.9793	.9799	.9805	.9811
79	.9816	.9822	.9827	.9833	.9838	.9843
80°	.9848	.9853	.9858	.9863	.9868	.9872
81	.9877	.9881	.9886	.9890	.9894	.9899
82	.9903	.9907	.9911	.9914	.9918	.9922
83	.9925	.9929	.9932	.9936	.9939	.9942
84	.9945	.9948	.9951	.9954	.9957	.9959
85°	.9962	.9964	.9967	.9969	.9971	.9974
86	.9976	.9978	.9980	.9981	.9983	.9985
87	.9986	.9988	.9989	.9990	.9992	.9993
88	.9994	.9995	.9996	.9997	.9997	.9998
89	.9998	.9999	.9999	1.0000	1.0000	1.0000
90°	1.0000					

Table V Cos x (x in degrees and minutes)

x	0'	10'	20'	30'	40'	50'
0°	1.0000	1.0000	1.0000	1.0000	.9999	.9999
1	.9998	.9998	.9997	.9997	.9996	.9995
2	.9994	.9993	.9992	.9990	.9989	.9988
3	.9986	.9985	.9983	.9981	.9980	.9978
4	.9976	.9974	.9971	.9969	.9967	.9964
5°	.9962	.9959	.9957	.9954	.9951	.9948
6	.9945	.9942	.9939	.9936	.9932	.9929
7	.9925	.9922	.9918	.9914	.9911	.9907
8	.9903	.9899	.9894	.9890	.9886	.9881
9	.9877	.9872	.9868	.9863	.9858	.9853
10°	.9848	.9843	.9838	.9833	.9827	.9822
11	.9816	.9811	.9805	.9799	.9793	.9787
12	.9781	.9775	.9769	.9763	.9757	.9750
13	.9744	.9737	.9730	.9724	.9717	.9710
14	.9703	.9696	.9689	.9681	.9674	.9667
15°	.9659	.9652	.9644	.9636	.9628	.9621
16	.9613	.9605	.9596	.9588	.9580	.9572
17	.9563	.9555	.9546	.9537	.9528	.9520
18	.9511	.9502	.9492	.9483	.9474	.9465
19	.9455	.9446	.9436	.9426	.9417	.9407

x	0'	10'	20'	30'	40'	50'
45°	.7071	.7050	.7030	.7009	.6988	.6967
46	.6947	.6926	.6905	.6884	.6862	.6841
47	.6820	.6799	.6777	.6756	.6734	.6713
48	.6691	.6670	.6648	.6626	.6604	.6583
49	.6561	.6539	.6517	.6494	.6472	.6450
50°	.6428	.6406	.6383	.6361	.6338	.6316
51	.6293	.6271	.6248	.6225	.6202	.6180
52	.6157	.6134	.6111	.6088	.6065	.6041
53	.6018	.5995	.5972	.5948	.5925	.5901
54	.5878	.5854	.5831	.5807	.5783	.5760
55°	.5736	.5712	.5688	.5664	.5640	.5616
56	.5592	.5568	.5544	.5519	.5495	.5471
57	.5446	.5422	.5398	.5373	.5348	.5324
58	.5299	.5275	.5250	.5225	.5200	.5175
59	.5150	.5125	.5100	.5075	.5050	.5025
60°	.5000	.4975	.4950	.4924	.4899	.4874
61	.4848	.4823	.4797	.4772	.4746	.4720
62	.4695	.4669	.4643	.4617	.4592	.4566
63	.4540	.4514	.4488	.4462	.4436	.4410
64	.4384	.4358	.4331	.4305	.4279	.4253

TABLE V Cos x (x in degrees and minutes) 135

x	0′	10′	20′	30′	40′	50′
65°	.4226	.4200	.4173	.4147	.4120	.4094
66	.4067	.4041	.4014	.3987	.3961	.3934
67	.3907	.3881	.3854	.3827	.3800	.3773
68	.3746	.3719	.3692	.3665	.3638	.3611
69	.3584	.3557	.3529	.3502	.3475	.3448
70°	.3420	.3393	.3365	.3338	.3311	.3283
71	.3256	.3228	.3201	.3173	.3145	.3118
72	.3090	.3062	.3035	.3007	.2979	.2952
73	.2924	.2896	.2868	.2840	.2812	.2784
74	.2756	.2728	.2700	.2672	.2644	.2616
75°	.2588	.2560	.2532	.2504	.2476	.2447
76	.2419	.2391	.2363	.2334	.2306	.2278
77	.2250	.2221	.2193	.2164	.2136	.2108
78	.2079	.2051	.2022	.1994	.1965	.1937
79	.1908	.1880	.1851	.1822	.1794	.1765
80°	.1736	.1708	.1679	.1650	.1622	.1593
81	.1564	.1536	.1507	.1478	.1449	.1421
82	.1392	.1363	.1334	.1305	.1276	.1248
83	.1219	.1190	.1161	.1132	.1103	.1074
84	.1045	.1016	.0987	.0958	.0929	.0901
85°	.0872	.0843	.0814	.0785	.0756	.0727
86	.0698	.0669	.0640	.0610	.0581	.0552
87	.0523	.0494	.0465	.0436	.0407	.0378
88	.0349	.0320	.0291	.0262	.0233	.0204
89	.0175	.0145	.0116	.0087	.0058	.0029
90°	.0000					

x	0′	10′	20′	30′	40′	50′
20°	.9397	.9387	.9377	.9367	.9356	.9346
21	.9336	.9325	.9315	.9304	.9293	.9283
22	.9272	.9261	.9250	.9239	.9228	.9216
23	.9205	.9194	.9182	.9171	.9159	.9147
24	.9135	.9124	.9112	.9100	.9088	.9075
25°	.9063	.9051	.9038	.9026	.9013	.9001
26	.8988	.8975	.8962	.8949	.8936	.8923
27	.8910	.8897	.8884	.8870	.8857	.8843
28	.8829	.8816	.8802	.8788	.8774	.8760
29	.8746	.8732	.8718	.8704	.8689	.8675
30°	.8660	.8646	.8631	.8616	.8601	.8587
31	.8572	.8557	.8542	.8526	.8511	.8496
32	.8480	.8465	.8450	.8434	.8418	.8403
33	.8387	.8371	.8355	.8339	.8323	.8307
34	.8290	.8274	.8258	.8241	.8225	.8208
35°	.8192	.8175	.8158	.8141	.8124	.8107
36	.8090	.8073	.8056	.8039	.8021	.8004
37	.7986	.7969	.7951	.7934	.7916	.7898
38	.7880	.7862	.7844	.7826	.7808	.7790
39	.7771	.7753	.7735	.7716	.7698	.7679
40°	.7660	.7642	.7623	.7604	.7585	.7566
41	.7547	.7528	.7509	.7490	.7470	.7451
42	.7431	.7412	.7392	.7373	.7353	.7333
43	.7314	.7294	.7274	.7254	.7234	.7214
44	.7193	.7173	.7153	.7133	.7112	.7092
45°	.7071	.7050	.7030	.7009	.6988	.6967

Table VI Tan x (x in degrees and minutes)

x	0'	10'	20'	30'	40'	50'
45°	1.0000	1.0058	1.0117	1.0176	1.0235	1.0295
46	1.0355	1.0416	1.0477	1.0538	1.0599	1.0661
47	1.0724	1.0786	1.0850	1.0913	1.0977	1.1041
48	1.1106	1.1171	1.1237	1.1303	1.1369	1.1436
49	1.1504	1.1571	1.1640	1.1708	1.1778	1.1847
50°	1.1918	1.1988	1.2059	1.2131	1.2203	1.2276
51	1.2349	1.2423	1.2497	1.2572	1.2647	1.2723
52	1.2799	1.2876	1.2954	1.3032	1.3111	1.3190
53	1.3270	1.3351	1.3432	1.3514	1.3597	1.3680
54	1.3764	1.3848	1.3934	1.4019	1.4106	1.4193
55°	1.4281	1.4370	1.4460	1.4550	1.4641	1.4733
56	1.4826	1.4919	1.5013	1.5108	1.5204	1.5301
57	1.5399	1.5497	1.5597	1.5697	1.5798	1.5900
58	1.6003	1.6107	1.6212	1.6319	1.6426	1.6534
59	1.6643	1.6753	1.6864	1.6977	1.7090	1.7205
60°	1.7321	1.7437	1.7556	1.7675	1.7796	1.7917
61	1.8040	1.8165	1.8291	1.8418	1.8546	1.8676
62	1.8807	1.8940	1.9074	1.9210	1.9347	1.9486
63	1.9626	1.9768	1.9912	2.0057	2.0204	2.0353
64	2.0503	2.0655	2.0809	2.0965	2.1123	2.1283

x	0'	10'	20'	30'	40'	50'
0°	.0000	.0029	.0058	.0087	.0116	.0145
1	.0175	.0204	.0233	.0262	.0291	.0320
2	.0349	.0378	.0407	.0437	.0466	.0495
3	.0524	.0553	.0582	.0612	.0641	.0670
4	.0699	.0729	.0758	.0787	.0816	.0846
5°	.0875	.0904	.0934	.0963	.0992	.1022
6	.1051	.1080	.1110	.1139	.1169	.1198
7	.1228	.1257	.1287	.1317	.1346	.1376
8	.1405	.1435	.1465	.1495	.1524	.1554
9	.1584	.1614	.1644	.1673	.1703	.1733
10°	.1763	.1793	.1823	.1853	.1883	.1914
11	.1944	.1974	.2004	.2035	.2065	.2095
12	.2126	.2156	.2186	.2217	.2247	.2278
13	.2309	.2339	.2370	.2401	.2432	.2462
14	.2493	.2524	.2555	.2586	.2617	.2648
15°	.2679	.2711	.2742	.2773	.2805	.2836
16	.2867	.2899	.2931	.2962	.2994	.3026
17	.3057	.3089	.3121	.3153	.3185	.3217
18	.3249	.3281	.3314	.3346	.3378	.3411
19	.3443	.3476	.3508	.3541	.3574	.3607

Table VI Tan x (x in degrees and minutes) 137

	0′	10′	20′	30′	40′	50′
65°	2.1445	2.1609	2.1775	2.1943	2.2118	2.2286
66	2.2460	2.2637	2.2817	2.2998	2.3183	2.3369
67	2.3559	2.3750	2.3945	2.4142	2.4342	2.4545
68	2.4751	2.4960	2.5172	2.5386	2.5605	2.5826
69	2.6051	2.6279	2.6511	2.6746	2.6985	2.7228
70°	2.7475	2.7725	2.7980	2.8239	2.8502	2.8770
71	2.9042	2.9319	2.9600	2.9887	3.0178	3.0475
72	3.0777	3.1084	3.1397	3.1716	3.2041	3.2371
73	3.2709	3.3052	3.3402	3.3759	3.4124	3.4495
74	3.4874	3.5261	3.5656	3.6059	3.6470	3.6891
75°	3.7321	3.7760	3.8208	3.8667	3.9136	3.9617
76	4.0108	4.0611	4.1126	4.1653	4.2193	4.2747
77	4.3315	4.3897	4.4494	4.5107	4.5736	4.6382
78	4.7046	4.7729	4.8430	4.9152	4.9894	5.0658
79	5.1446	5.2257	5.3093	5.3955	5.4845	5.5764
80°	5.6713	5.7694	5.8708	5.9758	6.0844	6.1970
81	6.3138	6.4348	6.5606	6.6912	6.8269	6.9682
82	7.1154	7.2687	7.4287	7.5958	7.7704	7.9530
83	8.1443	8.3450	8.5555	8.7769	9.0098	9.2553
84	9.5144	9.7882	10.078	10.385	10.712	11.059
85°	11.430	11.826	12.251	12.706	13.197	13.727
86	14.301	14.924	15.605	16.350	17.169	18.075
87	19.081	20.206	21.470	22.904	24.542	26.432
88	28.636	31.242	34.368	38.188	42.964	49.104
89	57.290	68.750	85.940	114.59	171.89	343.77
90°	∞					

	0′	10′	20′	30′	40′	50′
20°	.3640	.3673	.3706	.3739	.3772	.3805
21	.3839	.3872	.3906	.3939	.3973	.4006
22	.4040	.4074	.4108	.4142	.4176	.4210
23	.4245	.4279	.4314	.4348	.4383	.4417
24	.4452	.4487	.4522	.4557	.4592	.4628
25°	.4663	.4699	.4734	.4770	.4806	.4841
26	.4877	.4913	.4950	.4986	.5022	.5059
27	.5095	.5132	.5169	.5206	.5243	.5280
28	.5317	.5354	.5392	.5430	.5467	.5505
29	.5543	.5581	.5619	.5658	.5696	.5735
30°	.5774	.5812	.5851	.5890	.5930	.5969
31	.6009	.6048	.6088	.6128	.6168	.6208
32	.6249	.6289	.6330	.6371	.6412	.6453
33	.6494	.6536	.6577	.6619	.6661	.6703
34	.6745	.6787	.6830	.6873	.6916	.6959
35°	.7002	.7046	.7089	.7133	.7177	.7221
36	.7265	.7310	.7355	.7400	.7445	.7490
37	.7536	.7581	.7627	.7673	.7720	.7766
38	.7813	.7860	.7907	.7954	.8002	.8050
39	.8098	.8146	.8195	.8243	.8292	.8342
40°	.8391	.8441	.8491	.8541	.8591	.8642
41	.8693	.8744	.8796	.8847	.8899	.8952
42	.9004	.9057	.9110	.9163	.9217	.9271
43	.9325	.9380	.9435	.9490	.9545	.9601
44	.9657	.9713	.9770	.9827	.9884	.9942
45°	1.0000	1.0058	1.0117	1.0176	1.0235	1.0295

Table VII Natural or Naperian Logarithms $\log_e x$ or $\ln x$

x	0	1	2	3	4	5	6	7	8	9
1.0	.00000	.00995	.01980	.02956	.03922	.04879	.05827	.06766	.07696	.08618
1.1	.09531	.10436	.11333	.12222	.13103	.13976	.14842	.15700	.16551	.17395
1.2	.18232	.19062	.19885	.20701	.21511	.22314	.23111	.23902	.24686	.25464
1.3	.26236	.27003	.27763	.28518	.29267	.30010	.30748	.31481	.32208	.32930
1.4	.33647	.34359	.35066	.35767	.36464	.37156	.37844	.38526	.39204	.39878
1.5	.40547	.41211	.41871	.42527	.43178	.43825	.44469	.45108	.45742	.46373
1.6	.47000	.47623	.48243	.48858	.49470	.50078	.50682	.51282	.51879	.52473
1.7	.53063	.53649	.54232	.54812	.55389	.55962	.56531	.57098	.57661	.58222
1.8	.58779	.59333	.59884	.60432	.60977	.61519	.62058	.62594	.63127	.63658
1.9	.64185	.64710	.65233	.65752	.66269	.66783	.67294	.67803	.68310	.68813
2.0	.69315	.69813	.70310	.70804	.71295	.71784	.72271	.72755	.73237	.73716
2.1	.74194	.74669	.75142	.75612	.76081	.76547	.77011	.77473	.77932	.78390
2.2	.78846	.79299	.79751	.80200	.80648	.81093	.81536	.81978	.82418	.82855
2.3	.83291	.83725	.84157	.84587	.85015	.85442	.85866	.86289	.86710	.87129
2.4	.87547	.87963	.88377	.88789	.89200	.89609	.90016	.90422	.90826	.91228
2.5	.91629	.92028	.92426	.92822	.93216	.93609	.94001	.94391	.94779	.95166
2.6	.95551	.95935	.96317	.96698	.97078	.97456	.97833	.98208	.98582	.98954
2.7	.99325	.99695	1.00063	1.00430	1.00796	1.01160	1.01523	1.01885	1.02245	1.02604
2.8	1.02962	1.03318	1.03674	1.04028	1.04380	1.04732	1.05082	1.05431	1.05779	1.06126
2.9	1.06471	1.06815	1.07158	1.07500	1.07841	1.08181	1.08519	1.08856	1.09192	1.09527

Table VII Natural or Naperian Logarithms log$_e$x or ln x 139

x	0	1	2	3	4	5	6	7	8	9
3.0	1.09861	1.10194	1.10526	1.10856	1.11186	1.11514	1.11841	1.12168	1.12493	1.12817
3.1	1.13140	1.13462	1.13783	1.14103	1.14422	1.14740	1.15057	1.15373	1.15688	1.16002
3.2	1.16315	1.16627	1.16938	1.17248	1.17557	1.17865	1.18173	1.18479	1.18784	1.19089
3.3	1.19392	1.19695	1.19996	1.20297	1.20597	1.20896	1.21194	1.21491	1.21788	1.22083
3.4	1.22378	1.22671	1.22964	1.23256	1.23547	1.23837	1.24127	1.24415	1.24703	1.24990
3.5	1.25276	1.25562	1.25846	1.26130	1.26413	1.26695	1.26976	1.27257	1.27536	1.27815
3.6	1.28093	1.28371	1.28647	1.28923	1.29198	1.29473	1.29746	1.30019	1.30291	1.30563
3.7	1.30833	1.31103	1.31372	1.31641	1.31909	1.32176	1.32442	1.32708	1.32972	1.33237
3.8	1.33500	1.33763	1.34025	1.34286	1.34547	1.34807	1.35067	1.35325	1.35584	1.35841
3.9	1.36098	1.36354	1.36609	1.36864	1.37118	1.37372	1.37624	1.37877	1.38128	1.38379
4.0	1.38629	1.38879	1.39128	1.39377	1.39624	1.39872	1.40118	1.40364	1.40610	1.40854
4.1	1.41099	1.41342	1.41585	1.41828	1.42070	1.42311	1.42552	1.42792	1.43031	1.43270
4.2	1.43508	1.43746	1.43984	1.44220	1.44456	1.44692	1.44927	1.45161	1.45395	1.45629
4.3	1.45862	1.46094	1.46326	1.46557	1.46787	1.47018	1.47247	1.47476	1.47705	1.47933
4.4	1.48160	1.48387	1.48614	1.48840	1.49065	1.49290	1.49515	1.49739	1.49962	1.50185
4.5	1.50408	1.50630	1.50851	1.51072	1.51293	1.51513	1.51732	1.51951	1.52170	1.52388
4.6	1.52606	1.52823	1.53039	1.53256	1.53471	1.53687	1.53902	1.54116	1.54330	1.54543
4.7	1.54756	1.54969	1.55181	1.55393	1.55604	1.55814	1.56025	1.56235	1.56444	1.56653
4.8	1.56862	1.57070	1.57277	1.57485	1.57691	1.57898	1.58104	1.58309	1.58515	1.58719
4.9	1.58924	1.59127	1.59331	1.59534	1.59737	1.59939	1.60141	1.60342	1.60543	1.60744

ln 10 = 2.30259	4 ln 10 = 9.21034	7 ln 10 = 16.11810
2 ln 10 = 4.60517	5 ln 10 = 11.51293	8 ln 10 = 18.42068
3 ln 10 = 6.90776	6 ln 10 = 13.81551	9 ln 10 = 20.72327

x	0	1	2	3	4	5	6	7	8	9
5.0	1.60944	1.61144	1.61343	1.61542	1.61741	1.61939	1.62137	1.62334	1.62531	1.62728
5.1	1.62924	1.63120	1.63315	1.63511	1.63705	1.63900	1.64094	1.64287	1.64481	1.64673
5.2	1.64866	1.65058	1.65250	1.65441	1.65632	1.65823	1.66013	1.66203	1.66393	1.66582
5.3	1.66771	1.66959	1.67147	1.67335	1.67523	1.67710	1.67896	1.68083	1.68269	1.68455
5.4	1.68640	1.68825	1.69010	1.69194	1.69378	1.69562	1.69745	1.69928	1.70111	1.70293
5.5	1.70475	1.70656	1.70838	1.71019	1.71199	1.71380	1.71560	1.71740	1.71919	1.72098
5.6	1.72277	1.72455	1.72633	1.72811	1.72988	1.73166	1.73342	1.73519	1.73695	1.73871
5.7	1.74047	1.74222	1.74397	1.74572	1.74746	1.74920	1.75094	1.75267	1.75440	1.75613
5.8	1.75786	1.75958	1.76130	1.76302	1.76473	1.76644	1.76815	1.76985	1.77156	1.77326
5.9	1.77495	1.77665	1.77834	1.78002	1.78171	1.78339	1.78507	1.78675	1.78842	1.79009
6.0	1.79176	1.79342	1.79509	1.79675	1.79840	1.80006	1.80171	1.80336	1.80500	1.80665
6.1	1.80829	1.80993	1.81156	1.81319	1.81482	1.81645	1.81808	1.81970	1.82132	1.82294
6.2	1.82455	1.82616	1.82777	1.82938	1.83098	1.83258	1.83418	1.83578	1.83737	1.83896
6.3	1.84055	1.84214	1.84372	1.84530	1.84688	1.84845	1.85003	1.85160	1.85317	1.85473
6.4	1.85630	1.85786	1.85942	1.86097	1.86253	1.86408	1.86563	1.86718	1.86872	1.87026
6.5	1.87180	1.87334	1.87487	1.87641	1.87794	1.87947	1.88099	1.88251	1.88403	1.88555
6.6	1.88707	1.88858	1.89010	1.89160	1.89311	1.89462	1.89612	1.89762	1.89912	1.90061
6.7	1.90211	1.90360	1.90509	1.90658	1.90806	1.90954	1.91102	1.91250	1.91398	1.91545
6.8	1.91692	1.91839	1.91986	1.92132	1.92279	1.92425	1.92571	1.92716	1.92862	1.93007
6.9	1.93152	1.93297	1.93442	1.93586	1.93730	1.93874	1.94018	1.94162	1.94305	1.94448
7.0	1.94591	1.94734	1.94876	1.95019	1.95161	1.95303	1.95445	1.95586	1.95727	1.95869
7.1	1.96009	1.96150	1.96291	1.96431	1.96571	1.96711	1.96851	1.96991	1.97130	1.97269

Table VII Natural or Naperian Logarithms $\log_e x$ or $\ln x$ 141

x	0	1	2	3	4	5	6	7	8	9
7.2	1.97408	1.97547	1.97685	1.97824	1.97962	1.98100	1.98238	1.98376	1.98513	1.98650
7.3	1.98787	1.98924	1.99061	1.99198	1.99334	1.99470	1.99606	1.99742	1.99877	2.00013
7.4	2.00148	2.00283	2.00418	2.00553	2.00687	2.00821	2.00956	2.01089	2.01223	2.01357
7.5	2.01490	2.01624	2.01757	2.01890	2.02022	2.02155	2.02287	2.02419	2.02551	2.02683
7.6	2.02815	2.02946	2.03078	2.03209	2.03340	2.03471	2.03601	2.03732	2.03862	2.03992
7.7	2.04122	2.04252	2.04381	2.04511	2.04640	2.04769	2.04898	2.05027	2.05156	2.05284
7.8	2.05412	2.05540	2.05668	2.05796	2.05924	2.06051	2.06179	2.06306	2.06433	2.06560
7.9	2.06686	2.06813	2.06939	2.07065	2.07191	2.07317	2.07443	2.07568	2.07694	2.07819
8.0	2.07944	2.08069	2.08194	2.08318	2.08443	2.08567	2.08691	2.08815	2.08939	2.09063
8.1	2.09186	2.09310	2.09433	2.09556	2.09679	2.09802	2.09924	2.10047	2.10169	2.10291
8.2	2.10413	2.10535	2.10657	2.10779	2.10900	2.11021	2.11142	2.11263	2.11384	2.11505
8.3	2.11626	2.11746	2.11866	2.11986	2.12106	2.12226	2.12346	2.12465	2.12585	2.12704
8.4	2.12823	2.12942	2.13061	2.13180	2.13298	2.13417	2.13535	2.13653	2.13771	2.13889
8.5	2.14007	2.14124	2.14242	2.14359	2.14476	2.14593	2.14710	2.14827	2.14943	2.15060
8.6	2.15176	2.15292	2.15409	2.15524	2.15640	2.15756	2.15871	2.15987	2.16102	2.16217
8.7	2.16332	2.16447	2.16562	2.16677	2.16791	2.16905	2.17020	2.17134	2.17248	2.17361
8.8	2.17475	2.17589	2.17702	2.17816	2.17929	2.18042	2.18155	2.18267	2.18380	2.18493
8.9	2.18605	2.18717	2.18830	2.18942	2.19054	2.19165	2.19277	2.19389	2.19500	2.19611
9.0	2.19722	2.19834	2.19944	2.20055	2.20166	2.20276	2.20387	2.20497	2.20607	2.20717
9.1	2.20827	2.20937	2.21047	2.21157	2.21266	2.21375	2.21485	2.21594	2.21703	2.21812
9.2	2.21920	2.22029	2.22138	2.22246	2.22354	2.22462	2.22570	2.22678	2.22786	2.22894
9.3	2.23001	2.23109	2.23216	2.23324	2.23431	2.23538	2.23645	2.23751	2.23858	2.23965
9.4	2.24071	2.24177	2.24284	2.24390	2.24496	2.24601	2.24707	2.24813	2.24918	2.25024
9.5	2.25129	2.25234	2.25339	2.25444	2.25549	2.25654	2.25759	2.25863	2.25968	2.26072
9.6	2.26176	2.26280	2.26384	2.26488	2.26592	2.26696	2.26799	2.26903	2.27006	2.27109
9.7	2.27213	2.27316	2.27419	2.27521	2.27624	2.27727	2.27829	2.27932	2.28034	2.28136
9.8	2.28238	2.28340	2.28442	2.28544	2.28646	2.28747	2.28849	2.28950	2.29051	2.29152
9.9	2.29253	2.29354	2.29455	2.29556	2.29657	2.29757	2.29858	2.29958	2.30058	2.30158

Table VIII Exponential Functions e^x

x	0	1	2	3	4	5	6	7	8	9
.0	1.0000	1.0101	1.0202	1.0305	1.0408	1.0513	1.0618	1.0725	1.0833	1.0942
.1	1.1052	1.1163	1.1275	1.1388	1.1503	1.1618	1.1735	1.1853	1.1972	1.2092
.2	1.2214	1.2337	1.2461	1.2586	1.2712	1.2840	1.2969	1.3100	1.3231	1.3364
.3	1.3499	1.3634	1.3771	1.3910	1.4049	1.4191	1.4333	1.4477	1.4623	1.4770
.4	1.4918	1.5068	1.5220	1.5373	1.5527	1.5683	1.5841	1.6000	1.6161	1.6323
.5	1.6487	1.6653	1.6820	1.6989	1.7160	1.7333	1.7507	1.7683	1.7860	1.8040
.6	1.8221	1.8404	1.8589	1.8776	1.8965	1.9155	1.9348	1.9542	1.9739	1.9937
.7	2.0138	2.0340	2.0544	2.0751	2.0959	2.1170	2.1383	2.1598	2.1815	2.2034
.8	2.2255	2.2479	2.2705	2.2933	2.3164	2.3396	2.3632	2.3869	2.4109	2.4351
.9	2.4596	2.4843	2.5093	2.5345	2.5600	2.5857	2.6117	2.6379	2.6645	2.6912
1.0	2.7183	2.7456	2.7732	2.8011	2.8292	2.8577	2.8864	2.9154	2.9447	2.9743
1.1	3.0042	3.0344	3.0649	3.0957	3.1268	3.1582	3.1899	3.2220	3.2544	3.2871
1.2	3.3201	3.3535	3.3872	3.4212	3.4556	3.4903	3.5254	3.5609	3.5966	3.6328
1.3	3.6693	3.7062	3.7434	3.7810	3.8190	3.8574	3.8962	3.9354	3.9749	4.0149
1.4	4.0552	4.0960	4.1371	4.1787	4.2207	4.2631	4.3060	4.3492	4.3929	4.4371
1.5	4.4817	4.5267	4.5722	4.6182	4.6646	4.7115	4.7588	4.8066	4.8550	4.9037
1.6	4.9530	5.0028	5.0531	5.1039	5.1552	5.2070	5.2593	5.3122	5.3656	5.4195
1.7	5.4739	5.5290	5.5845	5.6407	5.6973	5.7546	5.8124	5.8709	5.9299	5.9895
1.8	6.0496	6.1104	6.1719	6.2339	6.2965	6.3598	6.4237	6.4883	6.5535	6.6194
1.9	6.6859	6.7531	6.8210	6.8895	6.9588	7.0287	7.0993	7.1707	7.2427	7.3155

Table VIII Exponential Functions e^x 143

x	0	1	2	3	4	5	6	7	8	9
2.0	7.3891	7.4633	7.5383	7.6141	7.6906	7.7679	7.8460	7.9248	8.0045	8.0849
2.1	8.1662	8.2482	8.3311	8.4149	8.4994	8.5849	8.6711	8.7583	8.8463	8.9352
2.2	9.0250	9.1157	9.2073	9.2999	9.3933	9.4877	9.5831	9.6794	9.7767	9.8749
2.3	9.9742	10.074	10.176	10.278	10.381	10.486	10.591	10.697	10.805	10.913
2.4	11.023	11.134	11.246	11.359	11.473	11.588	11.705	11.822	11.941	12.061
2.5	12.182	12.305	12.429	12.554	12.680	12.807	12.936	13.066	13.197	13.330
2.6	13.464	13.599	13.736	13.874	14.013	14.154	14.296	14.440	14.585	14.732
2.7	14.880	15.029	15.180	15.333	15.487	15.643	15.800	15.959	16.119	16.281
2.8	16.445	16.610	16.777	16.945	17.116	17.288	17.462	17.637	17.814	17.993
2.9	18.174	18.357	18.541	18.728	18.916	19.106	19.298	19.492	19.688	19.886
3.0	20.086	20.287	20.491	20.697	20.905	21.115	21.328	21.542	21.758	21.977
3.1	22.198	22.421	22.646	22.874	23.104	23.336	23.571	23.807	24.047	24.288
3.2	24.533	24.779	25.028	25.280	25.534	25.790	26.050	26.311	26.576	26.843
3.3	27.113	27.385	27.660	27.938	28.219	28.503	28.789	29.079	29.371	29.666
3.4	29.964	30.265	30.569	30.877	31.187	31.500	31.817	32.137	32.460	32.786
3.5	33.115	33.448	33.784	34.124	34.467	34.813	35.163	35.517	35.874	36.234
3.6	36.598	36.966	37.338	37.713	38.092	38.475	38.861	39.252	39.646	40.045
3.7	40.447	40.854	41.264	41.679	42.098	42.521	42.948	43.380	43.816	44.256
3.8	44.701	45.150	45.604	46.063	46.525	46.993	47.465	47.942	48.424	48.911
3.9	49.402	49.899	50.400	50.907	51.419	51.935	52.457	52.985	53.517	54.055
4.	54.598	60.340	66.686	73.700	81.451	90.017	99.484	109.95	121.51	134.29
5.	148.41	164.02	181.27	200.34	221.41	244.69	270.43	298.87	330.30	365.04
6.	403.43	445.86	492.75	544.57	601.85	665.14	735.10	812.41	897.85	992.27
7.	1096.6	1212.0	1339.4	1480.3	1636.0	1808.0	1998.2	2208.3	2440.6	2697.3
8.	2981.0	3294.5	3641.0	4023.9	4447.1	4914.8	5431.7	6002.9	6634.2	7332.0
9.	8103.1	8955.3	9897.1	10938	12088	13360	14765	16318	18034	19930
10.	22026									

Index

CPSIA information can be obtained
at www.ICGtesting.com
Printed in the USA
FSHW022211150921
84783FS